IMPROVED KEELBOAT PERFORMANCE

A practical method of reducing transom drag.
Sabepe—Macwester No. 1. (*Denny Desoutter*)

Improved Keelboat Performance

FOX GEEN

Illustrations
by Maureen Verity

HOLLIS & CARTER
LONDON SYDNEY
TORONTO

TO SLIM

Paperback edition ISBN 0 370 10342 4
Hard-cover edition ISBN 0 370 10318 1
Printed and bound in Great Britain for
Hollis & Carter
an associate company of
The Bodley Head Ltd
9 Bow Street, London WC2E 7AL
by Northumberland Press Limited
Gateshead
Set in Linotype Baskerville
*First published as a paperback and
simultaneously as a hard-cover edition 1975*

CONTENTS

14
Other Motion, 178
*Heaving, 178 · Surging, 180 · Drifting, 180 · Harmonic
rolling, 181*

15
Performance Under Shortened Sail, 183

APPENDIX

Author's Foreword

This book may serve to close a gap in the vast range of sailing literature, which tends to run to extremes.

Learned books by the technically expert are hard to digest and often of small practical use to the sailor in search of bettering his sport. Alongside them lie volumes written by those who are highly knowledgeable and have much to contribute in narrow fields, such as ocean racing or dinghy sailing. These are usually clear and comprehensive within their limits, but it is a task to extract small bits of information from them which might be of use to you in your own field.

At the other extreme are books aimed at the novice or inexperienced sailor. A lot of these are apparently written for juveniles and can be condescending, which is an odd approach. Because a man is a little short on sailing know-how, it is irrational to regard him as dim-witted. He is only too often fully aware of problems connected with his boat and her uses. He seeks practical knowledge in assimilable form, preferably backed with the relevant theory so that he may be intellectually convinced of the correctness of his actions. Not only does he want to know 'how' but 'why'.

In the compass of these pages I have tried to condense as much knowledge as is needed by the silent majority of the sailing fraternity—those quiet folk who ask nothing

9

more than the ability to cruise successfully with family or friends. They take much pride in what they do and spend endless hours in pursuit of acceptable sailing performance. I am glad to be numbered among them.

A friend, reading my draft, said perceptively that I had written about the man-boat machine. I understood, but it was badly put. A man needs a boat to ensnare the fleeting winds and harness them for his pleasure; this does not make a boat a machine. A boat needs a man at the helm as she breasts the surly seas and tames them down; this does not make the man a machine. There is an empathy between them, and it is right that a man should care for his boat, and that she should look after him. So let it always be.

FOX GEEN
Weymouth, 1974

I

Introductory

The more you know about your boat the better you will be able to realise her full potential. Although the performance of any boat is governed by her original design, her owner can do much for her himself. You know how boats of the same type, even members of tightly restricted classes, vary considerably in their sailing abilities. Within such groups it is fairly easy to compare performances and set about improving the less successful contenders. That calls for attention to many details, large and small, which in the aggregate can transform a back-marker into a winner.

In the field of cruising such convenient comparison is seldom possible. Nevertheless, there are many contented owners who took the time and trouble to find out what was wrong with their boats, or was unsatisfactory about them, and put their findings to practical use. It is open to anyone to do the same.

Permanent improvement to performance can be made in a number of ways. A boat can be made to sail faster; easier to helm; less tender, or less stiff; more seakindly; more weatherly; less demanding on her crew; and many other things. Satisfactory performance can rest with a number of such matters, but you may justifiably consider that attention to a single item will resolve your problems. Perhaps it will, but after paying attention to it you may be tempted to delve deeper and find that more can be done with beneficial results.

Part of improvement will concern what might be called static changes; altering the shape of a rudder; adding or repositioning ballast; increasing sail area or the aspect ratio of a rig. This type of action will flow from an analysis of design and an understanding of certain fundamental principles.

The other side of improvement concerns the running of a boat; paying attention to proper sail setting and handling; balance; helming; trim; the fitting of efficient gear. This will stem from an understanding of principles and an habitual adherence to routines indicated by them.

The static considerations concern the boat as an inanimate object, a cadaver to be operated on. The others comprise a sort of empathy, a giving and taking between man and boat to get the best out of both.

In these pages I describe practical measures which will give certain results whether or not you bother about the underlying reasons. However, a background of knowledge is always helpful and to this end I have included a little theory in digestible form; you should not find it hard to swallow. It has been difficult to sort the wheat from the chaff within the compass of a small book like this. Only where it seemed to have been of advantage for practical purposes have I been discursive. In other instances you will find a bald statement of theoretical findings. These are as accurate and factual as possible; if you wish to plunge deeper into such turgid waters there is a list of recommended reading at the back.

Good performance means taking the boat as a whole, so that most of the factors are interrelated in one way or another. I plead this as a reason for not having written an outstandingly logical sequence of chapters, although in the end it will all have knitted together. Nor did it seem sensible to try and divide each aspect of the subject into two neat halves, one dealing with theory and one with practice. So that the text will flow smoothly and retain your interest, I have relegated calculations and such things to an appendix which you can refer to if you wish. There is no glossary of terms and abbreviations; these are introduced as they arise and this should not cause any confusion as there is a comprehensive index.

I would like to emphasise one point before getting down to details. Structural alterations are the *last* things to be thought about in a search for improved performance. Satisfaction will usually come from tending to running matters like tuning and balance. There can be enormous gains from just attending to sailing gear and the way it is handled or from looking at your boat's trim, the way she sits on and moves through the water.

It could be, of course, that in the final analysis you decide to add a skeg, alter the shape of the rudder, or even add a bowsprit and mizzen mast to your craft. There could be benefit in doing so, and many owners are continually trying major experiments to alter the characteristics of their boats. Some of these are mentioned as ways of looking at things, but this book confines itself to more orthodox methods of the kind that are within the competence of most of us. Having read, the choice of action is yours. I have endeavoured to avoid misleading words and figures but performance cannot be judged by absolute standards. My idea of perfection might be regarded as sloppy by one man and impossibly stringent by another; it is entirely a matter of opinion.

LINES AND PLANS

When you hear someone criticising the 'lines' of a boat he *may* know what he is talking about. He may not, because it is a very loosely used manner of expression and usually applied to the appearance of a boat and not to her design.

Designers formalise their idea of a boat by drawing a number of plans. These are accurately scaled and fully detailed so that a boat can be built from them without further communication between designer and builder. Such a procedure is of comparatively recent origin and for countless years boats were conceived and created on the basis of a model. The builder would make a scale model, or more usually a half-model, discuss it with the intended owner and by trimming it here and there and otherwise tinkering with it agree with him on a final model which had acceptable 'lines'. The boat would then be a replica on a full-size scale.

Good boats, and many of them, were built in such fashion and quite a number of them are still afloat today. You can take off the 'lines' of any boat and commit them to paper if the plans cannot be made available. It is not a difficult process, although fiddly. I once did this for my friend Arthur Stevens who, believe it or not, had the lead keel and deadwood removed from his boat when he was at sea about Her Majesty's affairs. From the measurements taken I could determine all the factors necessary to design a new keel and rudder for *Saga*. Fortunately she floated level on her marks when launched, for the materials had cost £200, and sails ably around Weymouth today.

Such work can be undertaken by anyone with the necessary ability to take accurate measurements and do simple sums. As the design of hull and sails is the most important aspect of any boat the next couple of chapters deal with the subject in detail.

Once you have learned to read plans (and it is not difficult) it will be pretty easy to deduce the probable characteristics and behaviour to be expected of the boat built from them. If you are not entirely happy with your own boat you should then be able to determine to what degree, if any, her faults are attributable to design.

Good designers can be relied on to draw plans from which pleasant and seaworthy boats can be built. Designing is more art than science and there is an element of luck in the matter, albeit small. Many reputable designers are only too pleased to forget about their early efforts, but they bear in mind the hard lessons learned in the former years. Like wine, they improve with age.

Regrettably, there are other designers whose products range from indifferent to downright poor. In these days of great demand for boats of all sorts, a number are marketed which are based on poor design, often amateur efforts, and may also be suspect in workmanship. Clever promotion and slick salesmanship may mask the reality, and if thinking of buying a new boat you would be wise to make more than a cursory enquiry into the stated abilities and advantages of new models. I do not contend that all amateur designers are incompetent as this would be far from the truth—there are many good ones like the late and incomparable Harrison Butler—but reputable moulders employ good, established designers and their products are to be relied on.

Notwithstanding this, I would hesitate to buy any boat unless I had been able to look at her plans, at least the lines drawings. There is a justifiable reluctance on the part of the designers of racing rule-cheaters to disclose their secrets, but a cruising boat needs no such security. A conscientious designer will always show and discuss plans with a genuine prospective buyer.

This brings out a point in case you intended to use plans for such purposes as modifying your own hull or sail plan. Designers live by selling their plans. They expect to receive a royalty from every boat taken from them, and this is no more than just. It is morally indefensible (also illegal) to misuse plans by, for instance, passing them on to someone else so that he can build from them. Play fair and keep them to yourself, especially the sheet entitled 'Table of offsets'; a hull can be built from this alone.

2

The Hull

I crave your indulgence, and patience, when you read through this chapter. It may seem that, because there is little that can be done to the structure of a hull to change its lines, you are being led away from practical matters. This is not so. The understanding of lines is of itself fascinating, but as you progress with the book you will find that there are many applications of the theory contained in this chapter: principally in the way of determining factors for calculations but also by a basic understanding of how water flows around a hull.

The key to the shape of a hull lies in the sheer plan, half-breadth plan and body plan (or plan of sections). These are usually set out on a single sheet and are collectively termed the lines plan, or lines drawings. You will some-times see the body plan superimposed on the sheer plan; in some ways this is useful but at first sight can seem a little confusing.

You can see that the sheer plan is a side view of the hull. The half-breadth plan is a view from below and divided longitudinally by a centreline (CL). One half is generally outlined by the edge of the deck and the other portrays 'diagonals'. The body plan is divided vertically

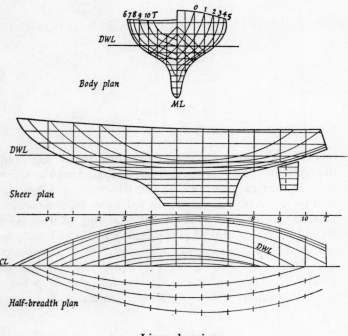

DWL

Body plan

ML

DWL

Sheer plan

CL

DWL

Half-breadth plan

1. Lines drawings

by a midline (ML), one half showing the view from forward and the other from astern.

Sets of lines are superimposed on the outlines and have the effect of slicing through the hull in three planes. Horizontal slices are referred to as waterlines, longitudinal vertical ones are buttock lines and thwartwise vertical ones are sections. These terms are not strictly accurate but are in common use and cause no confusion, so they are used throughout this book.

In reading lines you should grasp the fact that each

set of lines is common to all three plans but appears in different guises on each. This is obvious from the illustrations but can be stated like this:

Plan	Waterlines	Buttock lines	Sections
Sheer	Straight and parallel to designed waterline (DWL)	Curved	Straight and parallel vertically
Body	Straight and parallel to DWL	Straight and parallel to ML	Curved to match mid-section
Half-breadth	Curved to match DWL	Straight and parallel to CL	Straight at 90° to CL

On the body plans are lines radiating out from the ML diagonally, passing through all sections to end at the perimeter of the plan. They are diagonals and correspond to those shown on one half of the half-breadth plan. Unless he is actually building a hull they are of little concern to an owner. During construction he can check on the fairness of the hull by laying laths along the lines of the diagonals and comparing them with the plans. They have other uses which need not be mentioned here.

Once you have absorbed the intimate relationship between the three sets of lines you may find that if you gaze at the drawings, especially the half-breadth plan, the shape of the hull will materialise out of the paper. Another, more easily assimilable, way of portraying the shape of a hull is by means of an isometric drawing Fig. 2, but the lines are distorted and should never be used for taking measurements.

There is a convention which decrees that the bows shall be on the right-hand side of the drawings. It is

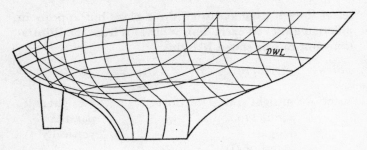

2. Isometric drawing

held that, once upon a time, the 'steer-board' was hung on what is now known as the starboard side and this gave it a certain cachet and precedence over port. This is open to question as models were used instead of plans until long after the steer-board had given way to the rudder. The convention is ignored by many designers because section lines are numbered on the plans starting from forward and it is easier and more logical to read them from left to right. Section lines are often referred to as 'stations', or in the case of the sheer plan, as 'ordinate stations'; you will see why later.

More often than not there will be eleven stations, starting with No. 0 at the fore end of the D W L and ending with No. 10 at its after end. This divides the D W L into ten equal portions and makes for simple calculation if metric terms are used, but life can be dreary if your D W L is 19 ft $6\frac{3}{4}$ in.

On occasion you will have to interpolate between a pair of lines on any of the three drawings, but this is easy. For instance, if your boat floats three inches above her D W L, you will want to know her actual load water-line (L W L) for finding out displacement and so on. Measure down the scale distance for three inches below the D W L and mark in a parallel line on both body and sheer plans. On the body plan step out with dividers or compasses from the M L to where the L W L cuts each section. At each station mark this distance out from the C L of the half-breadth plan. Join the marks with a curve faired to match the D W L and the W L immediately below it and you have drawn in, on all three plans, your own L W L.

You can similarly mark in additional sections; this often has to be done to lay off templates for bulkheads and so on. Draw in straight lines on the sheer and half-breadth plans; on the latter mark out to where they intersect W Ls and plot the resulting curve on the body plan. This can, of course, only be drawn in on one side of the body plan according to whether it lies fore or aft.

You are now in a position to use the lines plan for ascertaining facts about your boat, such things as her displacement, tons per inch immersion (t.p.i.), centres of buoyancy and lateral resistance, all of which are discussed in due course. The methods of doing so are set out in the Appendix.

There is a practical difficulty about accurately knowing where your L W L lies when you are in normal sailing trim, and this is critical information. If the D W L has been scribed on the hull, which is not often, and then has been obliterated with boot-topping or other paint, you cannot use it as a datum. Even if in evidence I would be inclined to ignore it, because quite often a D W L is wrongly marked in by the builder. Attack the problem empirically.

With your boat afloat in normal sailing trim, and in still water, get afloat in your tender so that you will not upset things by walking about the decks. Use a plumbline to measure the distance above water of the stem, centre of transom or counter, and a few other points along the rail which can accurately be transferred to the plans. Correctly scaled down, these will disclose your actual L W L; don't be worried if it is not quite parallel to the D W L. It is a matter easily corrected (see Chapter 10). The L W L should be the same distance from the D W L on both sides of the boat, naturally, and will only not be so if she has a list. This will be discernible.

That was a typical example of how you may often need to use full-scale measurements in preference to relying on a scaling up from the plans. These can be misleading. For one thing, paper 'moves'—shrinks and expands with changes of temperature and humidity. A large drawing could easily be an eighth of an inch different from one day to the next. At a scale of half an inch to a foot this would mean a discrepancy of three inches. Again, due to reproduction processes plans may not be exactly in conformity with the original drawings; and so on.

Always check the validity of scale measurements if you intend to use them as a basis for structural alterations.

The famous Admiral Lord Fisher propounded an equally famous dictum: 'The best scale for an experiment is twelve inches to the foot.' It is too true.

THE PLANIMETER

Areas can be measured mathematically, geometrically or with instruments.

It is easy to measure straight-sided figures and find their area by doing simple sums. Curved sides usually rule out such methods and math has to be used; where the curves are not of regular form—parabola, ellipse, and so on—this becomes a chore, and hull curves regrettably are irregular.

One practical but extremely laborious way of measuring

an irregular area is to transfer it to graph paper and count the little squares. By using the appropriate scale an area can be found within tolerable limits but the snag is in estimating the fractions of squares intersected by the contour. Decisions, decisions! ... There are other expedients but by far the best way of measuring is by using a planimeter.

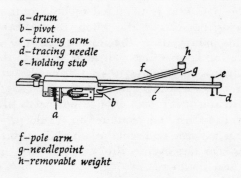

a – drum
b – pivot
c – tracing arm
d – tracing needle
e – holding stub

f – pole arm
g – needlepoint
h – removable weight

3. A planimeter

No designer could work without one, and if you are going to set about using your plans seriously for purposes of improving performance, you will have to use one. They are not cheap, but can sometimes be bought secondhand, or you could borrow one from a friendly draughtsman.

The instrument can be of fixed or of variable scale, the latter being considerably more expensive, although I picked up an excellent one for £3 a few years ago. A fixed-scale model is perfectly adequate for most purposes.

It consists of a pole arm and a tracing arm which are pivoted together. Near the hinge is a wheel having a calibrated drum around its edge. This abuts a vernier scale giving fine readings and leads through worm gearing to a coarse reading indicator. This is analogous to a sextant where you read large quantities from the scale and smaller ones from a vernier of one sort or another. One division on the coarse planimeter scale equals one hundred on the drum, each of which is further reducible on the vernier.

Each arm has a needlepoint under its extremity. The end of the pole arm is weighted to hold it into paper and drawing board and the whole instrument rotates around this point. Above the tracing needle is a small, freely rotating knob which you hold to guide the point along a line; a stub spaces it away from the surface of the paper.

The instrument is robust enough but needs to be used delicately. In following a contour you should press gently on the tracing arm to keep the wheel in contact with the paper. Do not overdo this or the wheel may slip and give false readings.

The fixed needle must be positioned outside the area to be measured and you should make a dummy run each time of use to ensure that the tracing needle can cover the whole contour without jamming.

When the wheel rotates it gives maximum readings for the distance travelled by the tracing needle and, when sliding at right-angles to its axis, gives minimum ones. When you start a trace, put the wheel at right-angles to the line at that point as this will minimise error caused when you try to stop tracing at the exact point at which you started.

The tracing should be in a clockwise direction and backtracking is to be avoided; it is better to start over again. If an area is too big to be measured in one go, divide it into sections, each of which has to be traced clockwise, and tot up the results.

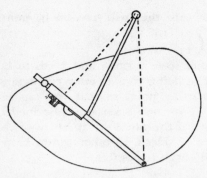

4. Position of planimeter when starting a trace

You will get the best results from fixed or variable-scale instruments if you calibrate them beforehand. Once this has been done there is no need to repeat the operation for a fixed-scale model, but beware of a slackened lock-screw on a variable-scale one. Draw an *accurately* measured and angled figure on a piece of drawing paper; you will know its area with certainty. Measure it several times with your planimeter and take the average of the readings; again, just as with a sextant. You can then construct a scale for future use. If, say, you measure 4 square inches and the average reading of your instrument is 14·84 divisions, 1 square inch will mean a reading of 3·71. This is then taken as a standard. 74·20 divisions would represent 20 square inches. If this were measured on a drawing scaled at $\frac{1}{2}$ in : 1 ft, the measured area would be 80 square feet in full scale. It all sounds complicated but is simple in practice.

You can allow for the 'movement' of paper by taking linear measurements before measuring areas. Suppose you discover that a designed 22 ft D W L has decreased to 21 ft 10 in. You can apply a correction to your readings to allow for this; they will have to be multiplied by a factor of $\left(\dfrac{21'\ 10''}{22'\ 0''}\right)^2 = 0.985$.

The vernier scale is small and needs to be read carefully, so when doing your dummy run arrange the stopping and starting point so that you can read without contortion. If you use a variable-scale model, error will be minimised if the wheel can rotate as often as possible for the contour traced. In other words, keep the adjustable arm as short as possible.

DERIVATION OF INFORMATION

Much can be deduced about the probable characteristics and behaviour of a boat from a study of her lines. Midsection 5A indicates that the boat will tend to be stiff and 5B that she will have opposite tendencies.

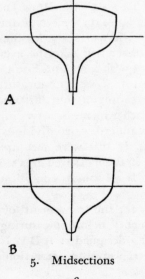

5. Midsections

Half-breadth 6A discloses a bluff-bowed hull with large wave-making potential which will be hard to drive and be slow in light airs. 6B has a fine entry and is recognisable as more racer than cruiser.

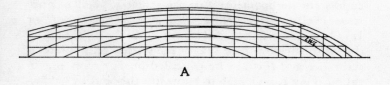

A

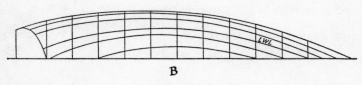

B

6. Waterlines

Before you can reach a stage of instant perception you will need to know about the principles concerned with and forces influencing a boat's behaviour. It is neither abstruse nor complicated, but mostly a matter of taking pains to measure accurately, calculate correctly and learn the terms used. You have met D W L, L W L, C L and others and it is time to progress.

Everything needed to help you in your labours is, I think, set out in the Appendix. Many readers will know much already and I beg their indulgence of what I include for the benefit of the uninitiate—as we all once were.

There are ridiculously easy ways of getting information which are used as a matter of course by the highly qualified. One of the loci frequently used is the centre of lateral resistance (CLR). Calculus and computer could track down this elusive point but I use a piece of cardboard and a pair of scissors.

Transfer to the cardboard the contour of the immersed part of the hull as shown on the sheer plan (omitting the rudder for the moment). Cut the shape out and balance it on a scissor-blade held at 90 degrees to the DWL, or LWL if yours is not coincident with the DWL. The line of the blade will be that of the vertical axis of lateral resistance and this can be pencilled in. For most practical purposes it is enough to know just this, but by balancing the card parallel with the WL you will find the longitudinal axis of lateral resistance. Where the two axes intersect is the position of the CLR.

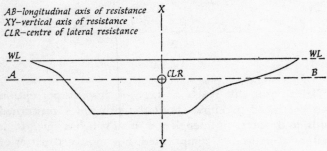

7. The centre of lateral resistance

In similar fashion, by using the LWL as a contour taken from the half-breadth plan you can locate the centre of flotation of the waterplane (CF).

Both CLR and CF lie at the centre of gravity (CG) of their respective planes. This coincidence is true for certain other loci but not all. The centre of pressure (CP) of a rudder is not at the CG of the immersed part of the blade. The centre of buoyancy (CB) is located within a volume, not an area.

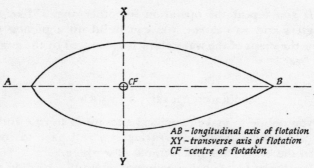

AB – *longitudinal axis of flotation*
XY – *transverse axis of flotation*
CF – *centre of flotation*

8. The centre of flotation

A boat's L W L changes shape as she moves in a sea-way, and affects her behaviour when under way. It is of interest to understand just how the W L changes with, say, different angles of heel and once more a simple method suffices.

If you draw mirror images of each half of the body plan you will end up with complete sets of fore and aft sections. On both drawings insert a new W L inclined at 15 degrees to the D W L, cutting it at the M L. Mirror out a half-breadth plan to be a full-breadth one and set out the new W L as previously explained. You will see that there has been a change of shape of the W L of this type:

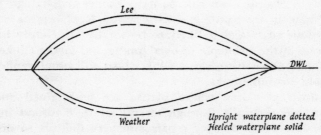

Lee

DWL

Weather   Upright *waterplane dotted*
                  Heeled waterplane solid

9. Effect of heeling on waterplane

If you repeat the operation for other angles, like 10 degrees and 20 degrees, you can build up a picture of how the shape of the waterplane is influenced by the angle of heel.

STRUCTURAL PROBLEMS

If you decide to make structural alterations to your hull, in addition to care in measuring you will have to consider the best means and methods of working.

Work can be of major or minor concern according to the materials of the hull, even if the task is identical. Take the simple case of removing a small section of hull for some reason, and making good afterwards.

With resinglass this can be minor. A section can be sawn out easily and repaired by lamination equally easily as long as you use proper moulding procedures. Even if the part of the hull worked on were highly stressed, efficient repair will put things back precisely as before.

A carvel plank, or section of planking, can be removed reasonably easily away from the timbers. As long as you pay attention to adequate securing, butt-strapping and other normal practices there should be little complication.

Clinker construction poses more problems. For a given length of hull there will be more frames and fixings; it is difficult to separate and sever planks. Replacement is a tedious affair and unlikely to be wholly satisfactory because of the principle of overlapping and faying clinker planks. Once disturbed, a clinker boat will usually suffer from incurable leaks.

Strip planking, where planks are both glued and fastened vertically through several layers, is almost impossible of satisfactory repair and very difficult to dismantle.

Moulded timber, of either hot or cold construction, can be sawn out as easily as can resinglass, but to repair the monolithic structure is a different matter. A replacement section will not bond with the original hull, of course, and such a patch will need to overlap considerably and be glued and screwed in place. It will stand proud internally and may have to be backed by some form of framing or shoring if highly stressed.

Conversely, to attach an A-bracket for a shaft to a wooden hull could be quite simple and very rigid, particularly if it were taken through a stout timber. A resinglass hull might have to be thickened up and reinforced to prevent cracking and flexion under power. This could be a very laborious task because of inaccessibility of the interior area concerned.

I will not run on, and just remark that it is best to think twice before cutting once. Sailors get into the habit of thinking like that, anyway.

Before leaving the hull for the time being, I would like to point out that resinglass and wood can be complementary to one another. The incorporation of wooden stiffening and reinforcing members into the relatively flexible material of a moulded hull is commonplace. There is no need to look askance at the practice and it will often provide a solution to many otherwise intractable problems.

There would seem no reason not to use the modern material to repair, stiffen, strengthen and otherwise prolong the life of a hull made of more traditional stuff. Hull and deck sheathing to stop marine attack and leakages is well known and widely used. Structurally, however, there seems always to be an inclination to replace like with like. This may not only be expensive but quite unnecessary. Instead of replacing a rotten frame with another piece of costly, almost unobtainable heart of oak

I would seriously consider using a resinglass channel moulded over a paper former but it would be essential to remove all traces of rotten wood beforehand. If properly laminated and meticulously bonded this could be even stronger than the original timber. Admittedly, it might not look very attractive, but this is a matter of balancing the aesthetic against the practical and there are many forms of camouflage.

To conclude, let me reiterate that structural alterations are a last resort.

3

The Sails

If the engine of your car, or powerboat, started to fall off
in performance you would notice the fact and do something about it. It is surprising how many sailing boats
are to be found with inefficient sails; perhaps this is
because they are inaudible. More probably, however, it is
because owners are not aware of the fact, especially cruising folk who are not seeking for ultimate speed. There
is much benefit in making a passage in as short a time
as possible for reasons of both safety and satisfaction.
Again, study of a sail plan can put you on the road to
better things.

THE SAIL PLAN

Fig. 10 is a typical sail plan; the amount of information
to be found on one is variable. Some of them have underwater lines drawn or dotted in, but this is inessential.
Those used for publicity purposes are usually sketchy
and display little more than the scale of the drawing—
sometimes not even that. A scale could be of the type
illustrated or just stated as $\frac{1}{2}$ in: 1 ft or, if metric, 1 : 15
or 1 : 20.

A decent working sail plan will be of slightly smaller
scale than the lines plan, but will still allow measurements to be taken off accurately and scaled up to full
size without difficulty. A designer who draws for amateurs

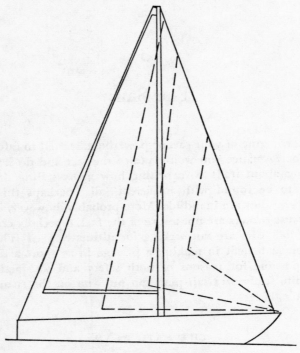

10. A typical sail plan

will usually include more detailed information than a professional builder would consider essential. Tables will give sail areas, dimensions of spars and rigging, the size of fittings and other useful information.

You can determine the precise location of chainplates, mast step, fairleads and tracking for sheets; measure mast rake and estimate sheet leads; and much else. The plan should enable you to put your boat afloat in sailing trim, although not necessarily fully tuned for efficient performance. The complete outfit of sails designed for the boat will be drawn in, and this is vital information.

THE CENTRE OF EFFORT (CE)

It is convenient to assume that a force acts at a point, centre or locus (which all mean the same thing). In the case of sails the loci are centres of effort and it is a simple matter to find the CE of a sail, or suit of sails (see Appendix page 201).

You can see that the plan shows sails fully stretched out along the CL of the boat, an impossible sailing situation. Each sail will have its CE and area marked in something like this:

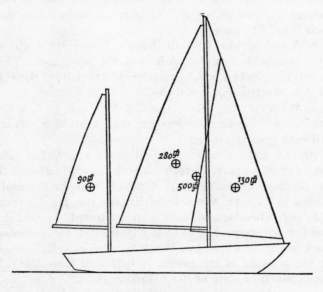

11.   Marking centres of effort and sail areas

and the combined C E of the working rig, usually main and No. 1 jib, is similarly marked in. These points are fictional, or theoretical, because in practice they are movable.

For aerodynamic reasons the C E of a sail moves forward as the angle α between sail chord and apparent wind decreases (see Appendix page 192). With α at a minimum the C E will in actuality be somewhere nearer the luff—about 35 per cent to 40 per cent of the chord abaft it, dependent on camber, which is considered later. The real and theoretical C E may coincide as far as the sail is concerned, but the true C E will *not* be located over the C L of the boat because the sail will be sheeted outboard. Aspect ratio (A R), as well as camber, is dealt with fully later, but in order that the full implications behind Figures 12, 13 and 14 may be understood, two facts should be known:

1. A sail of low A R will have a longer mean chord than one with a higher A R and the same area. This means that its C E will be relatively farther from the C L as it is sheeted more outboard.

2. When on the wind, the C E of a sail with a flat camber will lie further forward than that of a similar sail with greater camber.

At 12A a sloop is close-hauled in light airs. The true centre of effort ($C E_t$) is forward and slightly outboard of the designed centre of effort ($C E_d$). At 12B she is steeply heeled in stronger winds to which she can point closer; $C E_t$ is further forward and more outboard. At 12C she has the stronger wind over the quarter; $C E_t$ is coming back along the chord but is still forward of $C E_d$ because of the position of the boom. It is farther from the C L than in 12B in spite of the reduction in heel.

In all these cases the combined $C E_t$ has been out to one side, but look at Fig. 13.

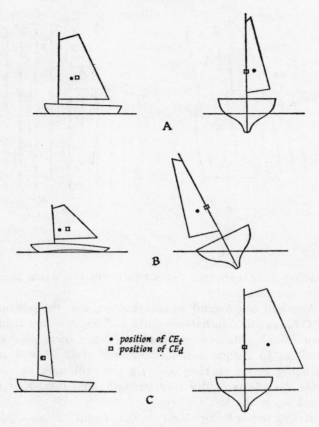

• position of CE_t
□ position of CE_d

12. Movement of CE_t

At 13A the foresail is blanketed and ineffective; only CE_{main} need be considered. As the boat bears away a couple more degrees, the foresail suddenly bellies out (13B) and sets on the side opposite to the mainsail, goose-winging. CE_{main} and CE_{fore} combine at CE_t which is then on or about the CL. At the same time the effective

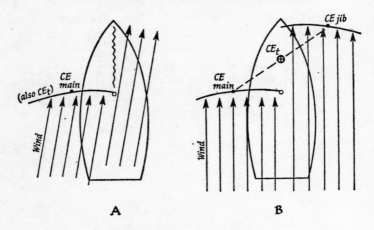

13. Effect of goosewinging on CE_t

driving force increases dramatically on the extra available sail area.

You will understand, as you read on, that the position of CE_t is critical to balance and handling. For now think how complex its movement will be when more sails are used, as in cutters and multi-masted craft. If you are thinking about altering your rig you will need to consider the effects of differing strengths and directions of wind upon the locus.

Rising winds bring about another complication.

With a properly designed suit of sails, CE_d will not change appreciably irrespective of which headsail is set with the *whole* mainsail. The area and C E of each sail will have been balanced out so that their effect on the position of CE_d is constant. Notwithstanding this, it should be obvious that CE_t *will* be affected by a change of headsail because CE_{genoa} can move much further outboard than $CE_{small\ jib}$. Changes to rig, as explained later, will entail using designed CE positions as a basis. In using them, you must always bear in mind the reality that it is, in practice, only the movement of CE_t that matters when sailing. This is very relevant when the question of reefing the mainsail comes into the question.

This action will lower its CE_t and cause it to move outboard to a lesser degree than when unreefed. Unlike changing a foresail, however, reefing causes a change to the position of CE_d, as shown:

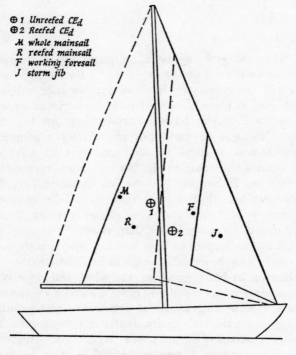

⊕ 1 Unreefed CE_d
⊕ 2 Reefed CE_d
 M whole mainsail
 R reefed mainsail
 F working foresail
 J storm jib

14. Centres of effort when reefed

It is of immense benefit to work out C E locations for your boat under different combinations of shortened sail. It may disclose that certain of these could put C E$_t$ in such a position as to dangerously unbalance your boat in bad conditions (see Chapter 15).

I omit all reference to spinnakers as they are not really suitable for cruising purposes, and need almost a whole book devoted to their management. There is a good exposition in *Race Your Boat Right* by Arthur Knapp Jr., who for many years was spinnaker-captain on the American 12-metres.

All other things being equal, a bigger sail will provide more drive than a smaller one, but a simple increase in area will not necessarily improve performance unless the bigger sail is identical in all its characteristics with the smaller one. If you have a one-off boat this may be a difficult thing to ascertain, but one which is a member of a class is easy to check on. You can rig your boat with sails from a sister-ship or ask her owner to try your sails on his boat. Either way will give an immediate standard of comparison. Both lots of sails should be of similar dimensions and differences in efficiency must thus be attributable to causes other than size.

You cannot expect to have any sail that is suitable for all conditions. Some will be good in light weather and deplorable in bad: good on the wind and poor off it. The converse can apply. Those which give tolerable results over a wide range of conditions and points of sailing are particularly desirable to a cruising owner. This applies especially to mainsails which are arduous to change under way. If you can afford to carry numerous headsails designed for specific purposes there is no problem. If not, choose general-purpose headsails for economy of both cash and effort.

Past a broad reach size becomes more important than other criteria, and with the wind aft is the predominant factor affecting drive.

ASPECT RATIO (AR)

Sails can look vastly different, ranging from the old-fashioned gaff to the tall, narrow modern mainsail and from low foresail with long foot to a high-clewed jib.

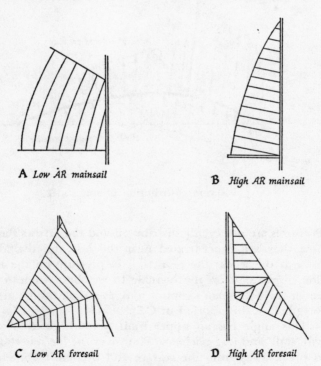

A *Low AR mainsail*

B *High AR mainsail*

C *Low AR foresail*

D *High AR foresail*

15. Aspect ratio

Each shape has its own characteristics, an important one being its A R (see Appendix page 194).

Air at rest exerts a constant atmospheric pressure, but when it flows over a sail it causes a reduction in pressure along the lee side and an increase on the weather side. The difference in pressure produces a total force F (see page 57), part of which is used to drive the boat.

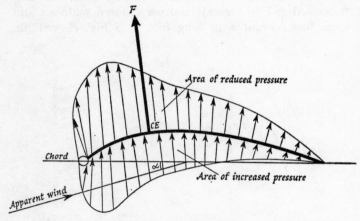

16. Pressure distribution about a sail

Pressures are not evenly distributed and the arrows show how they are concentrated near the luff and dwindle towards the leach; the reduction in pressure on the lee side is greater than the increase to weather. Angle α is under the helmsman's control and, as stated, can be used to influence the position of CE_t.

This angle has an upper limit beyond which a sail will 'stall' and lose efficiency. Flow over the lee side starts to break away from the surface and become turbulent; the low pressure starts to rise and so reduces the pressure differential. With a rigid aircraft aerofoil (wing) the stall angle is a critical point beyond which the wing is virtually useless. This is not so with a sail and the stalling effect is spread over a range of angles α which are greatly dependent on A R. A sail of A R6 stalls badly once angle α exceeds 12 to 14 degrees, but with A R1 the sail remains

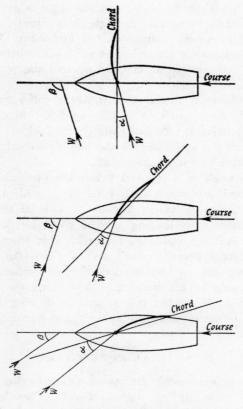

17. Maintaining optimum angle

reasonably efficient up to 35 to 40 degrees. A clipper ship sail of $AR\frac{1}{2}$ or $\frac{1}{3}$ would not stall fully until about 45 degrees.

For every sail there is an optimum angle α which will give maximum drive. As Fig. 17 shows, such an angle can be maintained by correct sheeting despite changes in the angle β between the boat's heading and the apparent wind (see Appendix page 192).

Once the boom fouls the shrouds, angle α will start to increase at the same rate as angle β, eventually reaching the angle of commencement of stall. For sails of AR6 this will be somewhere around $70° + 12° = 82°$, whereas a sail of AR1 will stay quite efficient up to some $70° + 35° = 105°$.

Beyond these angles the pressure differential will gradually drop until, with the wind aft, most of the available drive comes from the direct incidence of the airflow on the after side of the sail. Obviously, this is where sheer size starts to matter.

When angle α is reduced below the optimum, a sail is 'pinched' and starts to lose efficiency and eventually reaches the point where it starts to lose its shape and collapse along its leading edge, a condition known as 'luffing'. AR has nothing to do with this phenomenon, which occurs because a sail is not rigid like a wing. This time there is interference with the airflow on the weather side by the mast, and flow along the lee side strikes directly on the sail instead of flowing along it. This equalises pressure on both sides and causes luffing.

PRACTICAL APPLICATION

Take two boats with the same sail area but of the different types at Figs. 15A and 15B. They both sail at optimum angle α, which may be different for each boat. The greatest zone of pressure differential is near the luff, as explained, and so the sail with the longer luff has more of its total area in that zone and produces more drive. This is another way of saying that higher A R is more efficient on the wind, and accounts for the increasing tendency of racing craft to increase the A R of their sails because of the proportion of time spent beating.

Off the wind the tale is different. Because of the delay in stalling and other reasons which will become apparent later, sails of low A R produce more drive off the wind for a given area. On a broad reach a sail of A R1 will provide one and a third times the power of one of A R3 and one and a half times that of an A R6. As the wind gets further aft the efficiency of the low A R sail will drop off more rapidly than others. Other effects of A R will arise in due course.

CONTOUR

The contour of a sail also has a bearing on its windward efficiency. The most efficient sail would be of elliptical form, something like the wing of the old Spitfire fighter. The bermudian shape is not the best and gunter sails, those set on high-peaked gaffs, and some ultra-modern ones with full-length battens and wide aerofoil section masts more nearly approximate the ideal.

Other considerations will arise, but in the light of what has been said so far, satisfactory performance for either racing or cruising could be tackled in one way or another.

For racing boats, use a high A R to give optimum windward performance. Consequential inefficiency off the wind can be offset by using special purpose headsails and changing them as frequently as dictated by conditions.

This calls for large crews and a very large number of sails.

Such a state of affairs is unacceptable to a placid cruising man, a shorthanded family sailor or a single-hander. Adoption of a low AR rig, say gaff or gunter, would minimise the need for frequent changes of head-sail especially if the boat were made cutter-headed with roller-furling foresails. Off-wind efficiency would be enhanced and many cruising men are content to stay away from a close-hauled course; if it is unavoidable, recourse can be had to motor-sailing. However, the simple addition of a main topsail makes an incredible difference to the windward potential of a gaffer by adding in sail area where it is most effective: in the line of luff and well up in regions of undisturbed air.

That was a fairly typical example of how theory can be put to practical ends in your search for better performance. It is also, perhaps, the place to point out that nearly all performance aspects are closely inter-related; when considering one modification, beware of taking it in isolation because it could possibly have detrimental effects in other directions. For instance, increasing AR might mean stepping a taller mast. This could be heavier and need stouter rigging, which would bring in the question of CG, ballasting and transverse stability; also structural matters like the strength of decking and chainplates.

CAMBER

Camber is the profile, round or belly initially given to a sail during the process of cutting and finishing. Although camber is a critical feature of every sail, it is not shown on the sail plan. A designer may envisage sails with certain characteristics for his creation, but camber is an element which can be chosen by an owner to suit his particular wishes. To a degree it is also controllable under sailing conditions; this aspect is dealt with in Chapter 7, but for now the text is concerned solely with theory.

The amount of camber is usually expressed as the fraction $\dfrac{depth}{chord}$ and it will later be seen that it may not be constant over the whole height of a sail.

The shape of camber is defined as the distance from the luff, along the chord, of the maximum depth of camber.

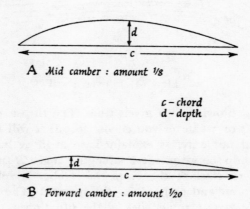

A *Mid camber : amount* ⅛

c – chord
d – depth

B *Forward camber : amount* 1/20

18. Shape and amount of camber

The purpose of camber is to give a sail an aerofoil section something like a very thin aircraft wing. The pressures shown in Fig. 16 are produced by the passage of air over the sail surfaces, as shown in Fig. 19.

19A is a high-lift, or high-thrust producing, type and 19B is a low-lift, or high-speed, form. This is discussed later, but for now it suffices to say that the efficiency of a sail is principally a matter of getting the best out of

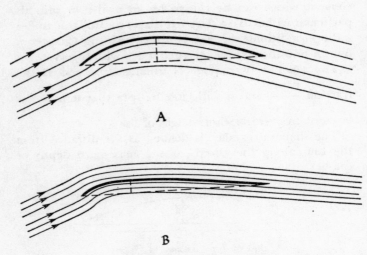

19. Flow of air over a sail

the wind blowing at a given time. The thrust, driving force, lift or whatever you choose to call it will be proportional not only, as explained, to angle α but to a relationship between wind strength, angle β and the amount of camber; the shape of camber is relatively unimportant and omitted for the moment. The following table gives a rough idea of the situation:

Angle β	Windstrength	Camber required
Up to about 30°	Light	Below 1/10
Up to about 30°	Strong	Above 1/20
30° to 75°	Light	1/10 to 1/8
30° to 75°	Strong	1/10 to 1/15
Over 75°	Light	⎫ Increasingly
	Strong	⎭ unimportant

48

Generally speaking, light winds call for more camber, and this should decrease as the wind rises. The further ahead the wind, the greater effect a difference in camber will have. With the wind aft it makes very little difference how much camber there is; a completely flat sail would be acceptable.

LOCATION AND INTERACTION

Impossibly but ideally sails should be totally unsupported—suspended in space—so that spars and rigging could not interfere with the airflow. In practice the best that can be done is to keep this interference to a minimum compatible with mechanical requirements. A foresail can be up to 50 per cent or more as efficient as a mainsail of the same area, and a mizzen sail proportionately less than that by about 30 per cent. This is because the mainsail is set behind a mast, and a mizzen is not only so set but in addition is unfavourably affected by the wind leaving the mainsail. Anything which causes eddies, turbulence or in any way disturbs an airflow and causes it to break away from close proximity to the cloth of a sail reduces efficiency. It is not possible to prevent a mast doing this sort of thing, and sizes and sections of masts are considered under the heading of rigging further along.

Aside from its intrinsic ability as a more effective type of sail, a foresail can also induce smooth airflow over the mainsail and delay the onset of stall. It is especially beneficial in the critical region of low pressure immediately aft of the lee side of the mast. This influence increases as angle β gets smaller, as shown:

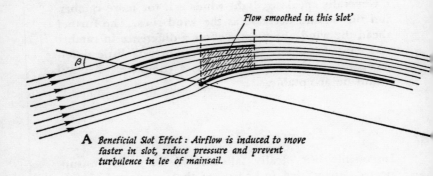

A *Beneficial Slot Effect: Airflow is induced to move faster in slot, reduce pressure and prevent turbulence in lee of mainsail.*

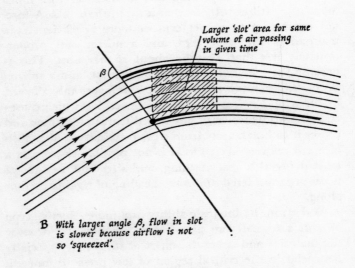

B *With larger angle β, flow in slot is slower because airflow is not so 'squeezed'.*

20. Slot effects

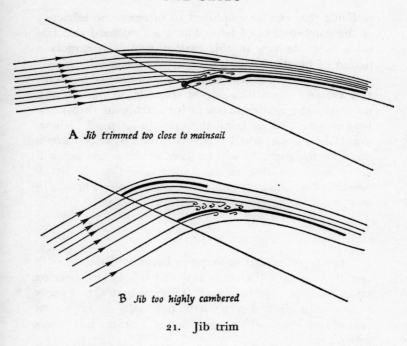

A *Jib trimmed too close to mainsail*

B *Jib too highly cambered*

21. Jib trim

Fig. 21 shows the importance of the airflow along the foresail. At 21A it has been positioned too close to the mainsail and at 21B has too much camber although positioned further outboard. In both cases the result is that the air impinges on the lee side of the mainsail, causing backwinding. This condition is undesirable because it disturbs the main airflow very early on; at worst it can render the mainsail almost completely inoperative. The effect can be almost entirely eliminated if the mainsail is fully battened from luff to leach, which keeps it in camber and permits an unobstructed flow along the sail. Full battening of mainsail, reduction of camber of foresail and adjustments to foresail sheeting are all expedients that can be employed to increase the efficiency of the combination of sails. There are no hard and fast rules to guide you in this matter, and it is largely a matter of practical trial and error, discussed fully later.

When you introduce mizzen sails, multiple headsails and esoteric items like mizzen staysails, flying jibs and watersails the matter becomes intractable in theory. As long as you bear in mind the need to keep all airflows over sails as undisturbed as possible, theory can be thrown out with the gash. As will be seen, at times one sail will be of more effectiveness than others, so that the maintenance of an efficient total airstream becomes a matter of boat handling as well as a static consideration.

CONDITION

If a sail is pulled out of shape or becomes distorted with age, its camber will change and this will have a bearing on its efficiency. Quite often it is for the better, and many racing men discard expensive, new sails in favour of threadbare ones which give better results. But don't rely on it.

A sail can be guilty of imperceptibly interfering with the airflow in a number of ways. Remedies are in your own hands. Changed camber can be rectified, quite often, by re-cutting or re-sewing a sail, whose shape was determined by the sailmaker as in Fig. 22.

At 22A the cloth of a mainsail is laid out flat, and the sail-plan shape is drawn in solid lines. The sailmaker actually cuts around the dotted lines, and the luff and foot are 'rounded'. They will nevertheless have to fit to mast and boom, which are straight, so that the extra area of cloth lying between the solid and dotted lines will enable the wind to belly out the sail, give it camber. In addition to this seams are sewn 'broad' as shown at 22B; they

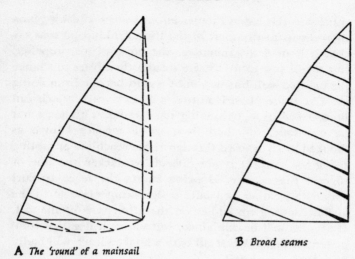

A *The 'round' of a mainsail*   **B** *Broad seams*

22. Cut of mainsail

take in more cloth at luff and leach. This also gives shape to the cloth, and a cunning combination of the two techniques enables a sailmaker to determine the amount *and* shape of the camber of the finished sail. By unpicking your sail you could adjust its features. Useful books on how to do this are listed at the back. Foresails call for slightly different treatment because they are not set on spars, but the principles are the same.

If changing the A R of a sail, you should remember that cutting any edge will automatically result in a re-positioning of camber and, if you change the shape of roundings, its amount as well.

An obnoxious failing of a sail is a fluttering or vibrating leach which grossly disturbs the slipstream and in the case of foresails can really upset the mainsail. If audible, this is known as 'motorboating' A clubmate of

53

mine sails the fastest Pionier in our waters, and his genoa motorboats madly most of the time. Unkind persons say that it is used to camouflage the noise of his propeller, but this is just calumny. He steadfastly refuses to change his beloved sail, but no doubt would benefit from doing so. The cause of such flutter is a slack leach which can be rectified by re-tensioning the tabling. This must not be overdone, for there is a worse menace known as 'hooked leach', caused through either leachline or tabling being too taut. It is not difficult to slacken the line or to ease the tabling. Hooking causes the leach to curl round to weather and offer a significant obstacle to the airflow, setting up eddies on the after parts of the sail. It also, as will be explained, introduces a negative component of drive. A foresail with a hooked leach will badly backwind a mainsail.

A thin sail will be porous and allow air to flow through the fabric from weather to lee, reducing the differential between high and low pressures.

A rough sail, which will result from sleeping on sailbags and otherwise misusing them, will increase the friction between air and cloth and slow down the flow.

Thick or uneven seams, even stitching standing too proud of synthetic cloth, will induce breakaway and turbulence. The further forward the fault, the more widespread will be its effect.

It is not advisable to fold sails if they are to be stowed in a bag or locker. They will eventually assume permanent creases having the same effects as other protrusions of the cloth. Roll them up without kinking the luff wire.

Vertically cut sails, such as gaff mainsails, offer a succession of seams to the crossflow of air and are relatively inefficient compared with those of horizontal cut. Their great advantage, which may outweigh their disadvantages for cruising purposes, is that if a seam lets go the sail will still remain partially effective, unlike the other type which will divide into two flailing halves and possibly disintegrate. Here again is a schism between the requirements of racing and of cruising.

4

The Wind

The kinetic energy of moving air masses is converted by the sails into a useful force. The manner in which this is done differs for three ranges of angle β, which I term conditions of sailing.

Condition 1: The air can flow freely over both sides of a sail. Angle α will be kept between the lower limit when the luff starts to lift and the upper one when the sail begins to stall. Range—from close-hauled to a beam reach.

Condition 2: Air can pass over both sides of the sail but not quite freely because angle α is over the upper limit due to cessation of boom travel. Range—from beam reach to quartering wind.

Condition 3: Air strikes at a large angle directly on the after side of the sail and flow over the lee side is extremely turbulent. Range—running.

There is a further situation within the range of Condition 1 where angle α is taken below the lower limit. Although air passes over both sides of the sail it produces little or no driving force; the sails luff or shake and the boat stops sailing.

Transition from 1 to 2 and back again will be conditioned by A R, being abrupt with a high ratio and gradual with a low one (see pages 43-44).

Transition from 2 to 3 and back again will be progressive and unremarkable.

In the following paragraphs a mainsail is taken in isolation, although practical considerations will make it necessary to take account of all types of sail and their combinations and interactions.

CONDITION 1

The airflow creates an area of low pressure to lee and of high pressure to weather of a sail. The total force (F) produced depends on sail area (S A), wind velocity (V) and the difference in pressures (dP).

The amount of camber will determine the pressures produced; generally, the greater the camber, the greater

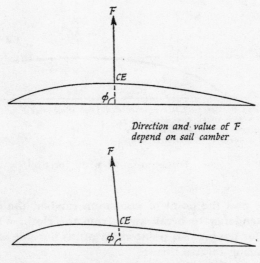

*Direction and value of F
depend on sail camber*

23. Effect of camber on wind force F

the pressures for given V, but other considerations must be brought in under actual sailing conditions.

As I said earlier on, F is assumed to act at the C E of a sail, but its direction of action depends on the shape of the camber. (Compare this with the profile of a rudder blade p. 157). The line C E-F is known as a 'vector', and indicates both a force and its direction (see Fig. 23).

REDUCED EFFICIENCY

In reality the wind acts along the whole length of a sail and dP varies from place to place. If any part of a sail is set inboard of a line parallel to the boat's course and tangential to the camber, an astern-acting component of F will exist at that position and act in opposition to forward thrust.

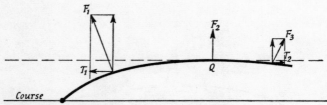

F_1, F_2 and F_3 are partial components of total wind force F
T_1 is forward-acting component of F_1
T_2 is astern-acting component of F_3
At the tangent Q, T is non-existent

24. Distribution of wind force

Once past the point of maximum camber, the airflow has a tendency to break away from the cloth. There is a case for having this point as far aft as possible without introducing reverse thrust.

Baffling of the airflow immediately aft of the mast, and

notably in the low-pressure lee, creates turbulence and diminishes efficiency. This can be greatly counteracted by the influence of the foresail over the main, the 'slot' effect illustrated in Fig. 20. Overlapping smooths out the airflow and the effect is enhanced if the point or area of maximum camber lies within the slot. Here again are conflicting requirements for different purposes.

A boat with foresails which are short in the foot, and so have little or no overlap on the main, will best be fitted with a mainsail whose camber lies well aft, perhaps a little aft of the middle of the chord. Many older boats have sails like this and seem to sail quite well with what would be, by today's standards, very baggy sails. Racing craft will have overlap no matter what headsail is set, except perhaps storm gear. This means that their mainsails will usually have camber well forward.

The matter comes to the fore when it is time to shorten sail in rising winds. The midget mainsails of racing craft are seldom reefed, and changes of headsail will reduce area adequately under all but extreme conditions. It is a different story with the cruising boat which has moderate fore canvas but must start to reef down at some stage in the proceedings.

A reduction to a sail without overlap destroys the slot effect, and a main with forward camber then loses a lot of its value at a time when it may be vital to keep the boat driving. A skipper may then tend to set too much sail forward and too little aft in an attempt to restore the slot. As explained later, this could upset balance to a dangerous degree. If you are thinking of buying storm jib and trysail, discuss them with the sailmaker. Trysail camber should be about central and moderately large as well, because under conditions of use you will be lucky to hold even a close reaching course. Flat sails are only of real value on the wind.

CONDITION 2

As angle β grows, turbulence increases, mainly over the lee side until with a quartering wind turbulence is gross. Efficiency drops off rapidly or gradually according to A R; sail area becomes of increasing significance.

At the beginning of the range, when on a beam reach, a boat is on her best point of sailing. Angle α is at an optimum so that the best is being got out of the airflow. Mutual interference between sails is very low although, paradoxically, so is such beneficial interaction as that of an overlap slot; forward camber is disadvantageous.

In addition to stalling effects getting worse, sails begin to interact inefficiently as the wind comes round to the quarter. After sails start to blanket forward ones, and a combination of large main and small fore sails suffers less than the converse.

CONDITION 3

Under running conditions most of the value of F comes from direct pressure on the surface of the sail lying at right angles to the wind. Area is critical and camber can be disregarded, but not A R.

Large eddies form behind leach and luff which detach themselves alternately and go downwind ahead of the boat. As each eddy leaves the sail there is a temporary decrease in pressure at that edge. This phenomenon causes the boat to roll out from side to side, 'rhythmic rolling'. This can be up to 60 degrees on either side in bad conditions and is most upsetting to a crew. There is little danger unless the wave period harmonises with the periodic rolling. In this case it will not damp out at all, but start to build up alarmingly. It is countered by a change of course and an alteration to sheeting. Experimental auxiliary sails designed to reduce rhythmic rolling have not been very effective. The size of eddies and pressure variations will depend on the lengths of luff and leach, so a high A R sail has a high rolling potential.

Goosewinging a foresail has beneficial effects on rhythmic rolling, but when this is feasible the mainsail can be 'by the lee'; the wind is at a critical angle and if the boat strays but a couple of degrees to leeward, the wind gets behind the main and gybes it. When running, especially in strong winds, it is sensible to lash the boom down to the lee rail to prevent an accidental gybe, which could cause damage and despondency. However, in big seas and with rhythmic roll present, the boom could hit the water with sufficient force to break it.

Unless you goosewing your foresail it will be so blanketed on a dead run as to be useless, but there is an expedient of value in light weather. Tack it down to the foot of the mast on the weather side. It will set well and provide extra drive. If you have a large foresail, and another of nearly comparable size, it may sometimes be even better to dowse the main and set both headsails on a forestay. In this way the C E is taken well forward and there is less imbalance than with two sails set centrally in the hull.

WIND COMPONENTS

Once more a single sail is taken to exemplify the way in which the available force F is broken down into its component parts, but in practice it will have effect at $C E_t$ for the complete set of sails in use (Fig. 25).

F can be resolved into a component acting to move a boat forward, T (thrust, drive, lift, propulsion), and one which acts at right angles to it, H itself (Fig. 26) can be broken into a lateral force tending to move the boat

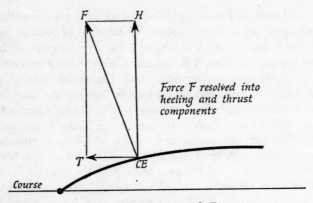

Force F resolved into heeling and thrust components

25. Components of F

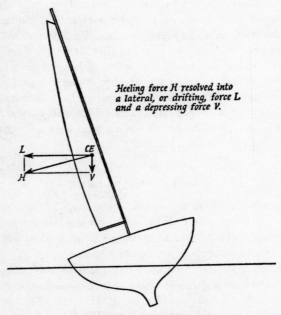

Heeling force H resolved into a lateral, or drifting, force L and a depressing force V.

26. Resolution of heeling force

sideways and another trying to push it downwards. The second of these is important to dinghies but not to keelboats.

For efficient performance you need to set and sheet your sails to produce maximum T and minimum H forces. Before going on to consider this, and other ways of getting the best out of the breeze, it will be useful to look at rigs and rigging.

5

Rigs and Rigging

Geographical, even local, circumstances play a part in deciding which rig is best suited to a particular area. When the prevailing wind blows on a lee shore and tidal streams run hard weatherliness, which is the ability to sail to windward, is important. Such conditions exist around British shores and here you will find a preference for sloops and cutters.

Off such places as the east coast of America, the wind is offshore and so on the beam of coasting craft for much of the time. There is a liking for schooners, ketches and other two-masted boats which can set clouds of sail between masts and have off-wind advantages.

Long-distance sailors like two masts. They confer a margin of safety in out-of-the-way places; make sail handling lighter work because the area is broken up into manageable portions; have handling abilities like self-steering and ease of heaving-to in bad conditions.

A boat which will steer herself for periods without either a hand on the helm or an ugly mass of machinery on the transom is highly desirable. I once owned a fine little sloop which was seakindly, able and all-round thoroughbred. Apart from lack of headroom, which nothing could be done about, she would not hold course for a moment without a hand on the tiller. Most well-behaved craft will round gently into the wind when the tiller is released, but she didn't and ran up or down according to the whim of the moment. This drove me to distraction until I fitted a mizzen mast and set eighteen square feet of sail on it. The boat was transformed. By balancing the mizzen against the jib I could get her to hold a steady windward course for long periods. Also, it made reduction of canvas a very simple matter. With genoa and mizzen alone she would make hull speed between Force 4 and Force 6, so that I got into the habit of lowering the main instead of reefing as the wind rose. It was nice to see the boom lashed out of harm's way, and heel reduced.

This kind of expedient can be adopted for all sorts of reasons apart from the ones given above. Most modern craft carry a very moderate amount of sail, insufficient to provide an adequate performance in light airs, and an addition to this can make all the difference between a dreary passage with a missed tide and six hours wait, and steady, satisfying progress.

A small mizzen mast, a bowsprit and extra headsail, a mizzen staysail, flying jib, watersail or other extravaganza can not only add to enjoyment but appreciably improve light-weather performance. The beauty of such things is that they are additions, not modifications, to existing rig and can be removed if necessary or left out of use in stronger winds.

Problems of weather helm are discussed on pp. 148-62 and one way of reducing it is to set extra fore canvas tacked to a bowsprit. This takes CE forward to an appreciable extent, but does not cause its height to increase, so that transverse stability remains unaffected. Many sailors regard a bowsprit as a nuisance in harbour and a danger at sea to foresail hands. Both these difficulties can be overcome by simple means.

Old-fashioned working boats usually had reefing bowsprits which could be run some way inboard when sail was shortened and completely for making fast along-side.

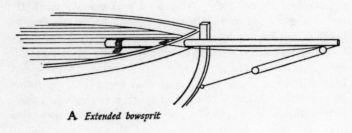

A *Extended bowsprit*

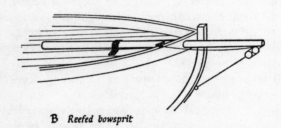

B *Reefed bowsprit*

27. A reefing bowsprit

The foredeck of a small, modern boat is likely to be restricted in length by the fore end of the coachroof, so that a sprit could not be fully reefed in this way. There are other alternatives to sliding it, and I once noticed a very workmanlike hinged sprit, shown in Fig. 28. The pin and its sockets should obviously be very stout, the fitment securely fastened through the deck and the bobstay arranged to be easily set up and slacked away. If the jib is not set flying (hoisted and bowsed down on

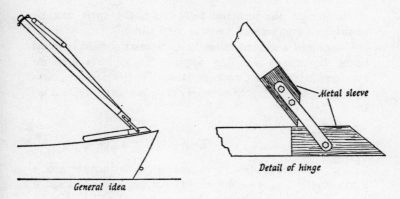

Metal sleeve

Detail of hinge

General idea

28. A folding bowsprit

its own luffwire), a forestay can be unshackled and secured out of the way at the foot of the mast. Tack, or tack and stay, can be hauled out to the fore end of the bowsprit with a traditional travelling ring fitted around the spar.

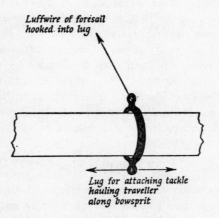

Luffwire of foresail hooked into lug

Lug for attaching tackle hauling traveller along bowsprit

29. A bowsprit traveller

If the hinge pin is made removable, the sprit can be unshipped and stowed away in bad times.

If you add a mizzen mast, neither mast should depend upon the other's standing rigging. Not only is such an arrangement difficult to tune decently, but separate sets of rigging give a factor of safety if either mast should go by the board.

ALTERNATIVE RIGS AND SAILS

Radical conversion from one type of rig to another might strike you as a way of improving performance, but space precludes anything approaching a full consideration of the subject. However, for the benefit of readers who might be thinking of fitting out, or rigging out, a D I Y hull I point out that many hulls are quite suitable to accept one of a number of rigs having different properties and abilities. I take it that few folk are ignorant of the general appearance and probable performance of sloops, cutters, yawls, ketches, schooners and of the differences between gaff, gunter, bermudian or other common type of sail configuration. Others exist, formerly in vogue on working and fishing boats, which have slipped into obscurity but are nonetheless worthy of close scrutiny.

Except for small, open craft the lugsail has fallen into disuse in the U.K. although quite a few are found elsewhere, notably in America where sailors are individualistic and unashamed to admit it. Varieties of lugsail are illustrated, and each has virtues and snags:

The dipping lug

This is probably the most powerful sail that has ever been devised, and has an obvious affinity with the sails of Viking and clipper ships of yore. With A R about

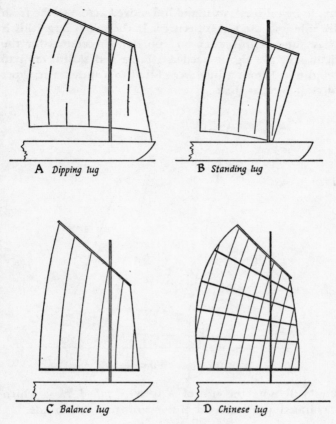

A *Dipping lug* B *Standing lug*

C *Balance lug* D *Chinese lug*

30. Types of lugsail

unity, it is efficient off the wind. As the airflow strikes a completely clear luff and is undisturbed by the mast, because the sail stands well clear of it at all times, it is efficient to windward as well. It is set well forward in the hull and you can see that its C E will be well aft of the mast. The tack is up in the bows and, when going about,

69

has to be released, yard and luff moved across to the fresh lee side and the tack resecured. If the sail is large this is heavy and skilful work and no doubt accounts for the demise of the rig; it needed a large and skilful crew to handle it. Dipping lugs were fitted to a number of types of craft such as this:

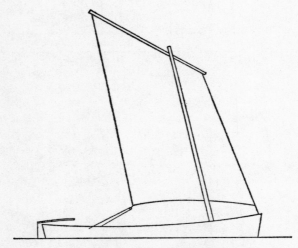

31. Scottish 25-ft. 'Fifie' fishing boat

You will note the use of a heavily raked mast which facilitated quick transfer of the yard from side to side.

The standing lug

The tack of this sail is taken to the foot of the mast and the yard always lies on the same side. It is inefficient when the yard is to weather and was normally only used as a mizzen sail. On the favourable tack it is almost as good as the dipping lug; it was, and still is, used for sailing tenders and other small, short-handed boats.

The balance lug

This rig combines the virtues of both standing and dipping lugs because it has a boom to which the sail is tacked and which automatically turns the luff to face the airflow. It is still in fashion for small dinghies like West Wight, Lymington, Solent and Burnham Scows but for larger craft it was disadvantageous, as noted by Dixon Kemp. The area of sail forward of the mast caused difficulties of staying due to airflow baffling. I am obliged to Dr Charles Milburn of Ontario for information on a class of balanced lug boat, the San Francisco Pelican, which is popular in America. In describing it he says:

'My interest in lugsails stems from owning one of these remarkable dinghies for several years. With their rig there is no problem in coming about, *because the mainsail does not extend much before the mast.* The rig is handy and all spars stow inside the boat. Performance-wise they are incredibly stable in heavy winds ...'

I use the italics to point out the fact that alteration to the original rig has eliminated its basic fault. Enquiring minds can often produce a solution to a problem, and there is no end to what can be tried in the interests of improving performance from one aspect or another.

Experimental work has gone on further. A larger version of the Pelican is fully battened and gives relatively much better performance in light airs than her original smaller sister design.

The Chinese lug

The Chinese lug, or junk sail, is a variation of the balance lug, although I should have put the horse before the cart as the junk sail is ancient. It is fully battened, like the larger Pelican, but in addition has a multipart mainsheet which leads to the after end of each batten, the number of which can be highly variable. This sail has been adapted with great success to set on Western hulls. The most renowned example is Colonel Hasler's *Jester*, which was raced across the Atlantic on several occasions. Commander King completed a circumnavigation in his *Galway Blazer*, a two-masted, junk-rigged ketch.

I am at present building a single-masted junk, quite unstayed, and cannot give personal account of experience under such rig. However, some years ago Thomas Colvin, the designer, builder and lug enthusiast of Virginia, U.S.A. experimented with three identical hulls. One was a bermudian ketch; one a gaff ketch with bermudian mizzen; and one a single-masted unstayed junk. The gaff rig had about ten per cent more sail area but otherwise the boats were statistically comparable.

Colvin established that all three were very similar performers to windward with the bermudian ketch having a slight edge in some conditions. Off the wind, however, both other boats were distinctly superior. From a broad reach to a run the junk boat left the gaffer well behind.

This bears out much of what has been said already in these pages. High A R has advantages on the wind, although the junk sail did not suffer unduly because of the minimal interference with the airflow. Off the wind low A R came into its own in providing extra drive, and the single sail scored because of lack of interference between sails, even though of only 90 per cent the S A of the gaffer.

THE SPRIT SAIL

This sail was the rig of the Thames trading barge and now of the Optimist dinghy, used to teach children how to sail.

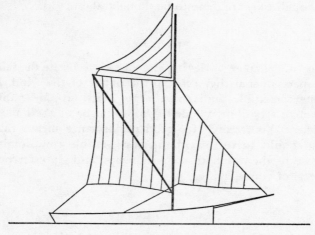

32.   Sprit sail

It is a type of gaff, but the sprit stands at the foot of the mast instead of the top. This keeps the C G of a heavy spar down without reducing its ability to spread the sail out to a maximum.

Reputedly, the sprit sail was simple to handle, and the crew of an 80-foot barge was usually no more than the skipper and a boy; between them they took the vessels on long coastal passages through all manner of weather. The bottom part of the sail can be brailed up quickly, giving a large reduction in area. If further reduction is needed, the sprit can be hauled closer to the mast and, if up against it, effectively dowses the sail.

Against advantages of handling, it is not a clean sail for windward work due to the interference with the airflow of large mast and sprit, but off the wind it has enormous driving potential.

There is an infinite variety of rig to be met with, but I have to stop short of such things as lateens and proas, although they are there to be thought about.

RIGGING

The sole purpose of rigging and spars is to locate the sails in space so that they can tap the power of the wind. If some incredible Indian Luffrope Trick would enable them to stay aloft unaided they would be at their most efficient. Your search for good performance means that you should strive toward the best possible compromise between the strength requirements and interference effects of your rigging.

MAST SECTION

The disturbance to an airflow rests with the size and shape of any object lying in its path, and the least resistance is offered by a streamline shape. A mast section of this type will only be effective as long as it points into the air stream, and when at right angles to it offers maximum resistance.

Some experimental boats have a rotating mast which lets the leading edge face the wind at all attitudes up to a beam reach. Usually the mast section is quite deep and the sail fits into a groove on its after edge; the sail is fully battened and the effect is to form a continuous aerofoil section.

The least objectionable profile for a general purpose mast is a circle, although for reasons of strength many

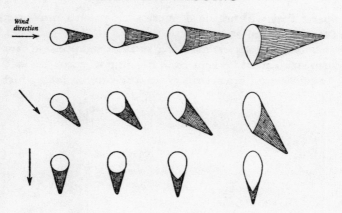

*Wind
direction*

33. Windshadow of mast sections

masts are of oval or pear-shaped form. If for any reason you wish to maximise efficiency in a particular attitude, say when beating, a specialised section will give the desired results but have disadvantages elsewhere. A round section means that the area of disturbed air stays fairly constant on all points of sailing and is therefore desirable for cruising.

The area of disturbed air will depend on the diameter of the section and its proportion to the length of the sail chord. Clearly, a thick mast will have a worse effect on a high A R sail than on a lower one (Fig. 34).

In seeking to keep your mast as thin as possible you start to run into a law of diminishing returns, because of the strength requirements.

MAST STRENGTH

Flexible and unstayed masts are used in narrow, specialised fields and I omit all reference to them because of space; they will not be of more than passing interest to the majority of sailors.

A normally stayed mast is a strut in compression, and such struts must be kept from bending when under load. Any deflection is a source of weakness and a mast which

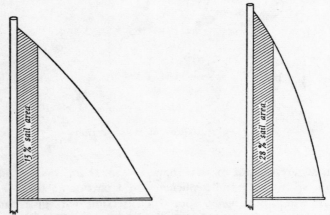

Greater percentage of High AR Sail lies in windshadow of mast.

34. Windshadow and AR

bends excessively will break or buckle. The compressive load depends on wind pressure, but the effect of this will vary according to the manner in which the mast is stayed.

Weight for weight, a tube will resist compressive loads much better than a rod, so that a stayed mast will usually be hollow. More resistance is obtained by increasing the diameter of the tube, but there is a practical lower limit to the thickness of the walls of a mast. Beyond it, even a slight deflection would load the walls to the point where they collapsed. Wooden masts have considerably thicker walls than metal ones, and the material is naturally more flexible, so that wooden masts may bend noticeably before giving way but metal ones suddenly buckle. Increased diameter means increased disturbance so that the requirements of strength and efficiency conflict. Compromise is reached by keeping diameter down as far as possible and combating the tendency to bend by means of staying the mast. As will be seen, the complex rigging called for with high A R and a long, thin mast poses problems.

Tautened wire vibrates, and rigging offers much more resistance to the airflow than its diameter would suggest. In closewinded conditions, this means that much of the air reaching the mainsail is already in a ragged state and a stage can be reached where complicated rigging starts to have a significantly detrimental effect on efficiency. As a general principle you should keep rigging as simple as possible, compatible with strength requirements.

It is not possible for me to specify sizes and strengths of rigging wire and fittings. These will depend so much on the individual boat and her intended purposes, that the only accurate guide will be the dimensions recommended by her designer. These may appear on the sail plan, but it is a good idea to find out what his views were if they do not. When boats are built down to a price, fitting-out yards may tend to skimp a bit in this area. For lengthy passages offshore, there is no harm in having slightly oversized wire and fittings. Not only will you have a safety margin, but rigging which is not taken close to its safe limits will be less stressed and last longer. For racing, of course, you may wish to balance integrity against performance, but it needs much consideration.

STYLE OF RIG

We live in the era of the masthead rig, where a foresail is set on a stay taken to the head of the mast. Older boats may have what is variously described as three-quarter, or seven-eighths, rig in which the forestay is taken to a point some distance short of this. Performance can, in most cases, be greatly improved by changing from old to new-style rig, but the design and purpose of both mast and rigging are essentially different in the two cases.

Whether one or two headsails are carried, the mast-head rig means that there is a greater percentage of total sail area forward of the mast. This puts a greater compression load on the mast and applies it higher up. The load put on the spar by the few upper square feet of the mainsail is relatively insignificant, and stresses only start to count when the mast takes the combined load of fore and main sails.

Accordingly, modern masts are of constant section throughout but older ones taper, sometimes steeply, above the hounds. They are equipped with spreaders and jack-stays designed to prevent the mast being pulled into an S-curve by the forestay, which is not opposed by the standing backstay. In some cases this opposition is provided by means of drifting backstays, attached at the hounds, which are an abominable nuisance. To stop obstruction to mainsail and boom, the weather stay has to be slacked away and the lee one set up taut on each occasion of changing tacks. If you omit, or fumble, this operation in heavy wind it can cost you your mast.

Years ago I was racing a 6-metre and the responsible hand wound in the lee runner far too tight in anticipation of rounding a mark. About a minute after hauling on the new board the mast imploded violently about six feet above the deck. There was a lot of despondency all round, and I learned the hard way the incredible compression strain that can be put on a mast by a stiff breeze.

A conversion to masthead rig does away with running backstays and jumper struts and jackstays, which gives much simpler, and less hazardous handling, and reduces windage of the rigging into the bargain. Also, more sail can be put forward of the mast, where it is more efficient. However, you will have to ponder on the implications of such a modification.

An increase of fore sail-area will take the combined C E forward and slightly upward, with effects on stability in the yawing and rolling planes. It will also increase compression on the top section of the mast to a great extent, so making a tapered mast basically unsuited to masthead rig. The positioning of crosstrees and shrouds will need to be looked at, as well as the strength of standing backstays and their fittings. It can be very expensive to buy a new mast, but some old ones are suitable for alteration; they will probably be of wood.

If such a mast is sound and its taper is not excessive, it can be shortened and then carefully slit down centrally to the commencement of taper, which will often be at the point of attachment of the lower ends of the jackstays. The mast should be throughbored for the insertion of a hardwood plug about $\frac{3}{4}$-in diameter at the bottom of the cut, glued in and left to set hard, to prevent splitting. A long wedge of suitable timber, shaped to fill out the mast to a constant section, can then be glued into the opened slit. An added precaution would then be to sheathe the mast with a layer or two of woven tape, glassed over the bared wood, from the truck down past the base of the join for about a foot.

The amount that you shorten your mast may mean cutting away a great part of the original taper, leaving little to be opened up and reinforced. The length detached will depend on the sail plan you have decided on, and in doing this remember that the mainsail will have to be shortened and that a foresail taken to the masthead will have both its luff and area considerably increased.

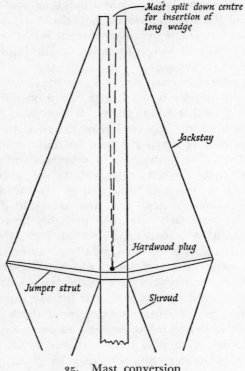

Mast split down centre
for insertion of
long wedge

Jackstay

Hardwood plug

Jumper strut

Shroud

35. Mast conversion

Also, the new configuration should be more efficient, so
that you may not need to carry as much sail as formerly,
so that the mast can be shortened quite a few feet.

The new angle of its luff will necessitate altering the
sheet leads for the foresail. It will, of course, not reach
masthead height and you might take it into use as a
No. 2 jib; a genoa of respectable proportions can be used
to give much valued drive to a converted boat.

To summarise, conversion from old rig to new is not

difficult, can be expensive or not according to your existing sails and mast, and will give much better performance.

CROSSTREES

Lower shrouds pull directly on the mast and need not be considered further.

Upper shrouds exist principally to prevent thwartwise distortion of the mast. In doing so they put a compression load upon it, and the angle they subtend with the mast is critical. It is determined by the length of the crosstrees. From the diagram you will see that the compression load is dramatically decreased if crosstrees are somewhat lengthened. The angle θ should never be less than about 12 degrees and the nearer to 15 degrees the better.

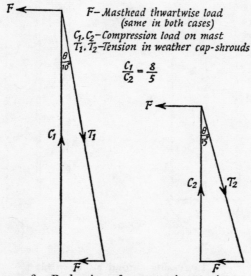

F – Masthead thwartwise load
(same in both cases)
C_1, C_2 – Compression load on mast
T_1, T_2 – Tension in weather cap-shrouds

$$\frac{C_1}{C_2} = \frac{8}{5}$$

36.   Reduction of compression strain

Crosstrees should divide the mast into roughly equal parts to distribute loads fairly. If they are about mid-height on a tall mast, increasing angle θ to 15 degrees may mean lengthening them until they will foul the leach of a large overlapping sail. One remedy would be to add a second pair of crosstrees and reposition the original ones, as shown:

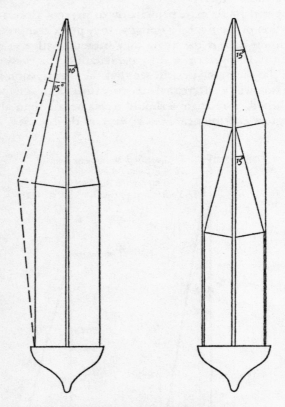

37. Increasing shroud angle without lengthening crosstrees

This would bring in a second set of upper shrouds with more windage and difficulty of tuning.

The more acceptable alternative for cruising would be to hollow out the upper part of the genoa leach, which is the sail most affected, so that fouling is minimised. As such a sail will not be sheeted hard in light airs and not set in stiff breezes, you may in actuality have no problem. When passage-making it is better to hold slightly off the wind and keep your boat tramping.

A crosstree is another strut in compression, and should preferably be hollow with thick walls. It is important to align crosstrees so that the angle they make with the shrouds is equal above and below their ends. If this is not so, they will be unfairly stressed and bend with predictable results.

It pays to have them of streamlined section to reduce interference with the airflow. Their junction with the mast is an obvious area of disturbance and end fittings should be small and neat. This is more feasible with a metal mast, when thin but robust stainless steel fittings can be pop-riveted on. Use professional size rivets of monel metal; the common or D I Y rivets are not likely to be man enough for the job.

FORE AND BACK STAYS

For ease of changing sail, which enhances crew performance, many boats are fitted with twin forestays. The idle one slackens as the foresail is bowsed down and the ill effects of thrumming are reduced or eliminated, although the airflow will be disturbed to some degree. The added security afforded to the mast is obvious.

There is a noticeable tendency nowadays for boats to be fitted with an inner forestay taken to the hounds in replacement of the forward pair of lower shrouds. This reduces windage and obstruction around the deck sides. It is debatable whether it affords adequate support to the lower part of the mast against the thwartwise forces imposed when a boat is on the wind. For cruising there is much to recommend the traditional arrangement, which also makes tuning simpler.

Twin backstays have everything to commend them. They give some added resistance to thwartwise forces and their baffling effect on an aft wind is negligible. Taken to each quarter they do not interfere with a self-steering vane, which can consequently be set central. A radar reflector can be strutted between them at an effective height. They afford an excellent factor of safety to the mast.

To get the backstay away from the centre of the transom rail, a single stay is often divided a few feet above deck level into two tails taken to the quarters. Not only does this remove the safety factor, but there can be up to five shackles in the rig instead of the single one used for a single backstay; each one is a point of potential weakness.

6

Setting Up the Rigging

The business of getting a boat to sail her best is often known as 'tuning', but this should not be taken to refer to the noise produced by highly tensioned rigging wire.

There is no purpose in over-tensioning rigging. Each wire will have to be individually set up according to its function, but once the point has been reached when the mast will stand without bending, in the circumstances to be described, further tightening will have bad effects.

Setting up the rigging puts an initial, and unavoidable, compression strain on the mast and crosstrees which is in the nature of things increased by wind forces and reaction to a seaway. Heedlessly winding away at bottle-screws will add appreciably to the load. There will be corresponding reactions at points of attachment to deck and mast and a resinglass hull, in particular, will distort under undue stresses. Rigging wire will be brought nearer to its safe working limit and its life will be shortened.

I previously mentioned that taut wire vibrates. Over-tensioned wire will vibrate excessively and can cause the metal to fatigue. This is especially noticeable with stainless steel rigging as the material, although stronger, is more brittle than mild steel. An indication of this is small 'needles' of wire, to be found on the deck, which have become detached from the stays.

MAST STEPPING

You will find three main types of mast stepping. Other masts exist which swing in lutchets, fold up, and for all I know are inflatable, but they are not worth talking about as they are confined to river and Broads craft.

The deck-stepped mast is deservedly popular, as it makes unstepping easy and, as it is shorter, is less expensive than others. The thrust is taken down to the keelson area through a pillar or similar device leading down from the deckhead beneath the step, or tabernacle. The mast is in effect pivoted at its step and has to be kept straight from that point upwards to the truck, even though it may then at times be wholly inclined away from the vertical.

Some designers favour the throughdeck, encastré, mast which rests on the keelson and is wedged into place where it enters the deck (partners). This kind of mast cannot pivot on its step and any deflection will impose a bending strain at the partners. It is extremely important to keep it straight throughout its length, and masts of this type are usually of stout dimensions to accept the extra rigging tension needed to stop them bending. They are more favoured in bigger craft where weight is not of such importance as in smaller ones.

A third type has to be watched for. It is a throughdeck mast unwedged and allowed movement within the deck aperture. It can pivot like the deck-stepped version and should be kept straight from heel to truck. Beware that you do not wedge or immobilise such a mast because it is likely not to be as stout as the true encastré mast. This kind of mast stepping is relatively rare, but was specified for the Royal Cape One-design—a variant of van de Stadt's famous *Black Soo*—one of the very few keelboats capable of real planing. The waters around

Cape Town do not offer much attraction to the died-in-the-wool cruising man, and the R C O D takes advantage of the southern rollers for exhilarating sailing over a distance. One boat has reliably been timed at 22 knots surfing.

SETTING UP THE RIGGING

If your boat has been out of commission, especially if hauled out, the initial setting up will only serve until the hull has sprung back to shape in the water and the boat has been sailed enough to take the stretch out of the wire. Coiled wire seems to get fractious in this respect; if you can possibly keep it straight, say by hanging it from a sail loft beam with a weight on the bottom, it will be of benefit when the time arrives to fit it. In any case, you will undoubtedly have to set up once more, but this should last a season unless you race enthusiastically or are out in a lot of poor conditions.

Thick wire will stretch less than thin for the same load, and cruising yachts will find advantages in it, greatly outweighing the slight extra cost, weight and resistance, not only in added security but in freedom from frequent adjustment.

Lee rigging will always slacken to some extent when a boat is hard on a stiff breeze, and this can be due to a combination of causes. Weather rigging will stretch to some extent and although your mast may be quite straight it may lean bodily away from the perpendicular to the deck; heavier gauge wire will, of course, stretch less. The mast may bend to leeward, usually above the crosstrees although the lower section of a thin mast may bulge to weatherward. This bend can be taken out by retensioning the upper (main) shrouds or using a heavier gauge wire for them. The material of a mast actually compresses under load and makes it fractionally shorter. This is inevitable and fortunately slackens the rigging with consequent reduction in compression load. It would be unwise to try to take out all lee slackness because you would impose intolerable tension on the mast and rigging.

SEQUENCE OF SETTING UP

There may be other ways of going about this business, but I think that logic makes it almost automatic.

Before all else, the mast has to be set upright; then it has to be raked to the required degree. After that, shrouds and stays are tensioned according to their disparate purposes.

THE MAST

If you can step your mast and initially set up the rigging ashore so much the better, but in many instances this has to be done with the boat afloat. Ashore, the craft can be levelled on its cradle or trailer and mast verticality checked with a builder's spirit level. A wedge tapered to the required angle can be put between mast and spirit level to check rake.

Afloat it is desirable, but not essential, to have a flat calm for the job. Belay the main halyard so that its eye touches a point marked on the rail roughly abeam of the mast step. If it can be brought across to touch a corresponding point on the other side, the mast will be vertical. Rake can be set in by hanging a weight on the halyard and using it as a plumbline, but the boat must obviously be still at the time. As you may have to adjust rake after sailing trials, I would not worry too much as long as it was small and not in a forward direction; this will be obvious if you view it from some distance abeam, say in the tender. Once you have finally settled on the angle of rake, you should check that the foot of the mast bears firmly all round on its step, inserting a wedge shim if needs be.

BOTTLESCREWS

The ends of bottlescrews are threaded in opposite ways. It saves frustration if you make sure that they are all fitted so that they can be adjusted by turning each one in the same direction for the same operation, say clockwise as you look down for tightening up.

LOWER SHROUDS

Verticality and rake are put in initially by using the lower shrouds, whether two pairs or one and an inner stay, as described. Other shrouds should for the moment remain slack enough to permit free movement of the mast. Give the lowers a fair amount of tension before going on to the other parts of the rigging; in this way they will hold the mast so that it will only bend, and not lean, above the point of their attachment. This is important with encastré masts. Once you have got the mast upright, bottlescrews should be adjusted in pairs, fore or aft, by giving an equal number of turns, or part-turns, to the bottlescrews, so that the lower shrouds have no thwart-wise effect.

MAST BEND

The only effective method of checking on this is to get your eye down as near the foot as possible and squiz up the track, or groove. It is made easier if the boom and gooseneck are unshipped. Even slight deflections in a fore and aft or thwartwise direction will be very obvious. Later, you will have to check on your mast under actual sailing conditions, and take sights from above the gooseneck with your pate pushed into the sail. This is awkward and tedious, but persevere with it; a glance around any congregation of boats will make it plain that many owners do not bother.

THE FORE AND BACK STAYS

The comparatively small angle it makes with the mast means that a forestay has to be highly tensioned. For good windward performance, the luff of the foresail should sag as little as possible, and this is set on the forestay. The weight put on the forestay of some offshore racers of moderate size is quite astonishing, in the order of tons, in an effort to achieve a straight luff. There are other means of reducing sag, duly dealt with in the next chapter, and I consider that a forestay should be taut but not excessively. If it is stressed to the limit, the hull will be wrung badly and you know how many racing craft are changed annually; an overstressed hull can lead to leaks and weaknesses in many directions.

One good reason for not allowing a forestay to sag too much is not often brought forward, but it is worth a mention: in hard, gusty conditions, often met with under the lee of high ground, a foresail will suddenly luff and then crack out like whiplash. A slack forestay allows it too much travel before bringing up, and a severe shock load is transmitted to the head of the mast. Shock loads can be several times the weight of a steady pressure of the same size, and a taut luff will minimise their effects by damping them down. The bigger the foresail the worse the shock, and a squall can shake your boat like a terrier with a rat.

This leads on to the realisation that large foresails need higher forestay tension, and is one of the self-defeating advantages of fitting high A R rig with small main and large headsails. All in all, the practical limit of a mainsail is about A R4, but even this means very large foresails. For cruising it is probably better to carry A R2 to 2.5 with correspondingly small headsails. With very low A R rig, such as some gaffers, there seems to be no need to tension the forestays very much at all, and they sail quite happily with jibs looking remarkably like opened sailbags: they are, admittedly, better off the wind.

The tension on the forestay, or twins if fitted, has to be opposed by the pull of the backstays if the mast is not to bend forward. As they subtend a greater angle with the masthead, the tension in backstays is considerably less than in their opponents, but fairly large. You will have to tighten down on forestay and backstay alternately, with glimpses up the track, to keep the mast straight until the required tension has been applied to the forestay.

In this process, it is quite possible that the mast, especially if wooden, will start to adopt a thwartwise curvature due to slight inequalities in wall thicknesses. This has to be taken out by tightening the main shrouds accordingly. If you have put enough tension into the lower shrouds, they will lose part of it as the mast shortens under strain, but this should not be enough to allow the mast to lean from the vertical.

This is debatable, but I confess a liking to have a slight aftward curve to my mast when the rigging has finally been set up. When fully pressed, the weight in the foresail takes this out, and I contend that it is at such times as this essential that the mast stand unbent.

The time of extreme loading on the mast is not when you are well shortened down in a yachtsman's gale of Force 6 or Force 7 but somewhere about the middle or top end of Force 4. It is then that a skipper holds on to full sail in the hope that conditions will ease a bit, and the odd gust ploughs the lee rail under and the boat staggers. At such a time the more support your mast has, and of the right sort, the better all round. Such things as this suggestion of leaving a little curve can, clearly, only be decided in the light of experience, and I do not recommend the practice otherwise. (See Chapter 15.)

UPPER SHROUDS

These provide resistance to thwartwise bending and should only be tightened to the point where they will

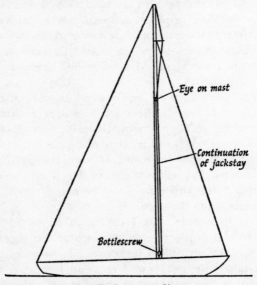

38. Jackstay endings

fulfil their function under sailing conditions. To allow the mast to lean and not bend, they should be of stronger wire than the lowers so that stretch is equalised for the longer and shorter lengths involved.

JACKSTAYS AND DRIFTING STAYS

Jackstays are a damn nuisance. They are meant to keep the top section of the mast straight by balancing the pull of the backstays at their top end and that of the forestay attached at the base of the jumper struts. In practice, unless assisted by drifting backstays, they do not seem to be very effective and need constant adjustment. If you are adorned with such a growth on your mast, it helps a lot to bring the stays down to terminate in bottlescrews at deck level; they can be led through eyes on the mast in place of the terminal fixings normally mounted.

Drifting (running) backstays should have a means of initially adjusting their tension like any other stay. If they have Highfield levers, these may have inbuilt means of adjustment; if not, a bottlescrew can be introduced into the lead.

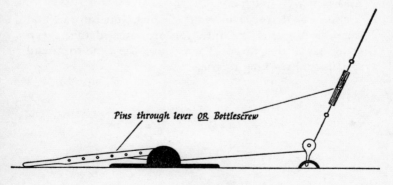

Pins through lever OR Bottlescrew

39. Adjustment of drifting backstays

Old boats often have their runners operated by means of block and tackle. As their foresails are small, the weight to be put into a runner is not great and running adjustment is simple under sail.

You may encounter examples of an obsolescent rig known unhappily as the 'slutter' rig. The well-known Vertue class is of the type. Light-weather headsails are set on a fixed topmast stay but others on a removable inner stay which is taken down from a point some distance below the truck and secured near the base of the fixed stay. No jumpers and jackstays are fitted but drifting backstays oppose the pull of the inner one. It is a seaworthy arrangement but unhandy.

The Vertue *Jamile* has crossed the Atlantic on several occasions under this rig, but her present owner has converted her to pure masthead rig. The original forestay was replaced by a heavier one and the backstay by a pair taken to the quarters. The bumpkin was removed, boom and mainsail foot shortened, but the slightly tapered mast was neither modified nor restepped. Her spars were extremely stout and Jack Sharpe thinks that there is no reason to doubt their suitability as they stand, although time will tell, no doubt.

Once you have set up your rigging, you can sail and carry out adjustments until you are satisfied that everything has settled down. Then comes the time to attend to sails and running rigging.

7

Making the Most of the Breeze

In Chapter 1 I observed that there is an interrelationship between all the facets of performance. When you are carrying out trials, no doubt several factors will be considered more or less simultaneously. For instance, mention of heeling will cause you to think about speed, transverse stability, drag, and other things like the effect of heeling on yaw and of rolling on pitch. It would not be sensible to cloud issues, and so I am taking each aspect affecting performance in isolation; you will have to link up associated matters appearing on other pages for yourself.

Before turning to less obvious ways of improving performance, it is crystal clear that you should ensure that your sails are providing maximum drive over the ranges in which you have an interest. (*Vide* Conditions 1, 2 and 3 in Chapter 4.) It may seem reasonable to suppose that improvement of sail efficiency in one direction should not have adverse effects in another, but this is not necessarily true. Unless you have means to enable sails to be used for varying conditions, they will have a good effect at one wind strength, or condition of sailing, and you will need to carry a selection of sails for others. For example, flat sails are good for sailing into hard headwinds but have inferior light-weather efficiency; such points as this will be examined in depth. In the end, you may well settle for moderately good performance over a wide range of conditions rather than for the peaks and troughs of specialised behaviour.

A subjective opinion, or feeling, that you have done something to improve performance is useless. You must have a valid standard of comparison. The dodge of trying out other boats' sails has been mentioned, but if nothing like this can be done, you have to use other means.

The most sensible criterion is that of speed through the water, and you will need a speedometer of one kind or another. There are other ways of observing changes in velocity like logging distance and elapsed time, but you really need a means of instantaneous indication.

HULL SPEED

Every displacement craft has a top speed above which she cannot be driven by her normal means of propulsion. This is her 'hull speed' (V_{max}), and the first move in sailing trials is to see how fast your boat will go. This may not be as fast as her potential V_{max} for reasons becoming apparent later, but for now suffice it to say that if she cannot get within acceptable distance of her potential speed she may be disproportionately slower at low speeds.

SPEED/LENGTH RATIO (R)

Theoretical hull speed for any boat is found from the equation

$$V_{max} = R \times \sqrt{L}$$

where V_{max} is in knots and L is the L W L in feet. R is a constant individual to any boat and dependent on her design characteristics. It should not, in theory, be above 1·34 (as explained later) but some exceptional and ex-

treme forms of hull have reached a speed equating to
R=1·45. This was because, in addition to almost per-
fect hull shape and finish, they had abnormal means of
propulsion. They were, in fact, grossly overcanvassed and
stressed beyond limits which would be acceptable to a
man wishing to preserve his boat. Broadly speaking, if
you have a racing boat you can be satisfied if R=1·275
and for a heavy-displacement cruiser R=1·2 would be
in order.

R can be found by simply inverting the above equation to read

$$R = \frac{V_{max}}{\sqrt{L}}$$

and you take V_{max} from the speed indicator and just measure your actual L W L. As the factor L is invariable for
set conditions, the value of R will vary directly as V_{max}.
The following table gives speeds attainable for various
lengths of W L over a range of values for R.

L W L (feet)	R=1·0	R=1·1	R=1·2	R=1·3	R=1·34
16	4·0	4·4	4·8	5·2	5·36
17	4·123	4·54	4·95	5·36	5·53
18	4·424	4·66	5·09	5·51	5·68
19	4·36	4·79	5·23	5·67	5·84
20	4·473	4·92	5·37	5·81	6·00
21	4·583	5·04	5·50	5·95	6·14
22	4·691	5·16	5·63	6·10	6·28
23	4·796	5·27	5·75	6·23	6·43
24	4·90	5·39	5·88	6·39	6·56
25	5·00	5·50	6·00	6·50	6·70

Speed in knots

These fractions of a knot may seem insignificant, but are not so at all. A cruiser of 20-ft L W L and R=1 will cover 107 miles on a 24-hour passage. If she can be cleaned up to sail at R=1·2 she will cover 129 miles in the same time. Putting it another way, the slower condition would mean taking 22 hours 21 minutes to cover 100 miles as against 18 hours 37 minutes in improved trim—a gain in time of 3 hours 44 minutes.

SAILING TRIALS

To find the maximum speed attainable by your boat as afloat you need a stiffish, steady breeze and fairly calm water. It helps if you can catch a breeze before it has been blowing long enough to set up a sizeable wave system. Seas introduce variable effects into the operation.

Fig. 25 shows how total wind force, F_t, can be resolved into driving and heeling components. When your boat has just attained maximum speed, which can be checked on the speedometer, any increase in F_t can only be converted into increased H; the craft will heel more and her speed should, it might seem, remain constant but in practice it might not.

When a boat heels, there can be two opposing effects. The shape of the waterplane changes and this can give rise to a reduction in speed (see Chapter 12). Conversely, heeling can lengthen the L W L which will allow a higher V_{max} to be reached i.e. the factor L in the above equation has been increased. This effect is appreciable if a boat has large overhangs and most inshore racing craft are designed to take advantage of the fact—the International metre classes, the old J-class, Dragons, 5·5 metres and others.

Part of your problem will be to minimise the detrimental effects of excessive heel. To demonstrate the result of unfairly balancing the T and H components of the wind-force, put your boat on a full-speed reach and then sheet in hard all round. She will slow down appreciably and heel dramatically.

This means that you can use the angle of heel, as well as speed, as an indicator of the effects of your adjustments and manipulations. It is easy to fit a simple clinometer in sight of the helmsman. If you can reduce the angle of heel without a corresponding reduction in speed you will have gained efficiency, and vice versa. Further, at speeds lower than V_{max}, if you can get more speed and less heel simultaneously, you are really getting somewhere.

A 'steady' wind usually remains constant in direction although its velocity fluctuates slightly above and below the mean. Due to the inertia of a moving hull these minor fluctuations have no significant effect on boat speed, but you must maintain the wind at a set angle of incidence (angle β, remember?—page 44). You can use the steering compass and be aided by a couple of sensitive telltales fitted to the lower shrouds. The weather telltale will be little affected by anything other than the prevailing wind.

Once you are sure that you have attained maximum available speed, make a note of the angle of heel, the speed and angle β.

You are then in a position to start experimenting, and need to be able to vary the tension of luff, leach and foot of sails, adjust sheeting and leads, alter the relative angle between sails, and try similar things all of which bring out the point that you must have efficient running rigging and fittings for the purpose.

Unless sails can be handled easily and efficiently you will never get the best out of boat or crew. All running gear and fittings should be stout, simple, easy to handle in all conditions and should, as far as possible, fail safe. Chandlers tempt with an abundance of shiny, expensive gadgets to which you should apply Occam's razor; the simplest solution to a problem is likely to be the best.

In light airs sails need to be set soft and sympathetic, with plenty of receptive curves, but in stiff breezes they must be bowsed down and flattened; this means applying considerable strain to them by way of the running gear. It is then that you will appreciate effective rigging and fittings. When it comes to a choice of gear, and ways of employing it, everyone will have his own ideas, methods and routines. What follows is merely suggestion, and one man's meat.

HALYARDS

Internal halyards, taken down through a mast and emerging at its foot, are fitted on many boats but I don't care for them for serious cruising. Racing men like them on grounds of reduced windage, but one minor disadvantage is that it is impossible to stop them rattling against the metal of a mast unless it is soundproofed. What I seriously deplore is that they are virtually impossible to re-reeve afloat if they part. Their sheaves are let into the mast and you would have to attach a block at masthead before you could rig an emergency halyard. As do other halyards, they wear about the sheaves. The length of halyard bearing over the top sheaves can be examined when sail is down but that at the foot of the mast is invisible when sail is lowered and partly so when it is hoisted. You can never be quite certain of condition.

MAINSAIL TRAVEL

The luff of a mainsail can travel in either a track

fitted to, or a groove designed in, the after face of the mast. Slides travel either internally or externally along the track; larger craft favour external slides which can be made of heavier stuff and are also less liable to jam. The drawing shows some of the ways by which slides can be fitted to eyelets let into the luff of the sail:

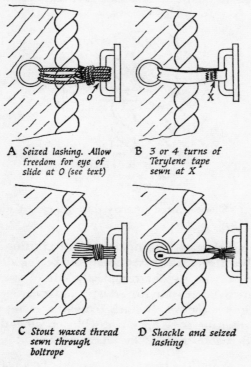

A *Seized lashing. Allow freedom for eye of slide at O (see text)*

B *3 or 4 turns of Terylene tape sewn at X*

C *Stout waxed thread sewn through boltrope*

D *Shackle and seized lashing*

40. Slide attachments

They may also be sewn into the luffrope. Sewn Terylene tape is neat and strong. Lashings have a nasty habit of chafing through or coming adrift. Shackles can distort, can scuff eyelets and cloth, wear out the slides, or wear through.

A jammed mainsail is hazardous at times and always a bother to free. Slides will be less liable to jam if the lashing or tape is not tightly bound to their eye and unable to move in one direction or another as strain comes on the halyard. Shackles have advantages in this respect.

If pin is too large to fit, it can be replaced by a lashing.

41. Shackle fitted directly through eye of slide

A grooved mast is unsuitable for a cruising boat. If the halyard gets adrift, say when shortening down, the luffrope is able to slip completely out of the groove. The sail balloons out to leeward and can be difficult to get under control. If track is used, a removable pin or a turnbutton at its open end will stop slides dropping out and enable them to be taken out one at a time as required.

Aerodynamically, grooved masting is preferable. Any space between luffrope and mast, inevitable with track, is eliminated. Air escapes from weather to lee through such a gap and reduces pressure differential in a critical region. Small vortices form along the whole length of the luff.

Some sophisticated (and costly) masts are extruded with both groove and track incorporated, which gives the best of both worlds:

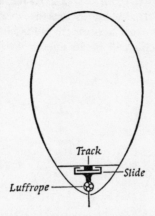

42.   Section of mast extrusion with integral groove and track

The sails of many gaffers and old-time craft are held to the mast by parrels, hoops taken right around the mast and secured to the sail luff in some way. Parrels cannot travel above the lowest shrouds, obviously, but these are usually in the region of the masthead on this type of boat. A gaping luff gap is to be seen, but a little more windward inefficiency on such craft is not of great moment.

BOOMS

A grooved boom has every advantage except that if points reefing is used instead of roller gear (and I approve of points reefing for offshore work), the points or lacing have to be taken under the boom instead of slipping between track and footrope. This is secure but unattractive.

At the foot of a mainsail (any sail, in fact) air forms vortices which curl from weather to lee, and cause drag and a drop in dP. A boom acts as a sort of endplate and diminishes vortex action; the wider it is across its top the more beneficial will be its effect.

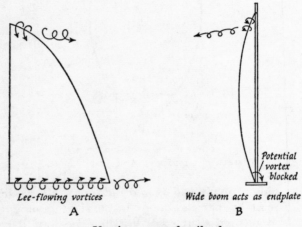

Lee-flowing vortices

A

Potential
vortex
blocked

Wide boom acts as endplate

B

43. Vortices around sail edges

As flexibility is desirable in some circumstances to be related, there is a case for making booms wider and less deep, but few examples are encountered, probably due to complications of reefing. A very wide boom, rigid, was tried out on *Enterprise,* the American 12-metre; it was known as the 'Park Avenue' boom. It carried short lengths of track fitted thwartwise at intervals on the top of the boom. Slides attached to the footrope travelled over these and could be clamped to give a set camber to the foot of the sail to suit given conditions:

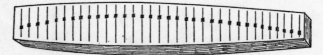

Camber along foot of mainsail controlled by position of slides

44. The 'Park Avenue' boom

This refinement gave better flow along the bottom part of the sail than would be possible if the boom were effectively rigid laterally. Some dinghy classes use a 'plank on edge' boom which flexes laterally and also gives a bit of unpenalised extra 'sail' area.

If you ever change from grooved spars to tracked, or vice versa, you will meet roping problems, as shown, which are however resolvable by a man handy with needle and palm (see Fig. 45).

Useful information on sailmaking and adjustment is to be found in many books, including the *Admiralty Sail-maker's Handbook*.

ADJUSTMENT TO CAMBER

Some sails, found more often on multihull boats, are fully battened from luff to leach and the bottom batten lies close to the top of the boom, to which the sail is only secured at tack and clew. The batten can be adjusted for conditions met with and permits the sail to be almost constantly cambered over its whole height. This can add five per cent, or more, to its efficiency at times.

The camber of normal sails is controllable within limits, although its principal characteristics depend on its cut. The basic method of controlling camber is by varying the tensions of luff, leach and foot. Luffs are tautened with the halyard, but stretch-luff sails are to be found where the roping is replaced by nylon tape. This

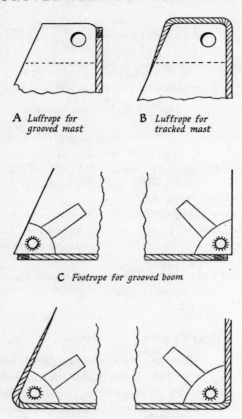

A *Luffrope for grooved mast*   B *Luffrope for tracked mast*

C *Footrope for grooved boom*

D *Footrope for tracked boom*

45. Variations in roping

allows the introduction of a 'Cunningham Hole', an eyelet set in the luff a foot or so from the tack. A small purchase draws this down towards the tack and gives a range of accurate control over the fullness of the sail in the luff area.

106

Luffs may need not only tensioning but positioning vertically on stay or mast, so that halyards should be complemented with a downhaul from the foresail tack to the base of its stay, and a downhaul from the inner end of the boom to the base of the mast. When fitting a downhaul, or indeed any other sort of tackle taken down to deck level, it is wise to bring the final lead up through an idle swivelblock. This lets you get pull by using the leg muscles, impossible if the strain is applied downwards.

The foot of a mainsail can be hauled out towards the end of the boom with a clew-haul, or small purchase of 2 : 1 or 3 : 1 advantage. The worm winches available are only suitable for large mainsails and will distort smaller ones if used. It is unnecessary to tension the foot of a sail excessively, and unless the creases which appear when the clew is hauled out disappear when the sail is well filled, tension is too great. The foot can also be tensioned toward the mast if a Cunningham-type eyelet is fitted a small way from the tack. In addition to reducing fullness, this brings the point of maximum camber forward, which is advantageous if the slot is present. Another way of considerably flattening the foot of the mainsail is by fitting it with a row of reefing points, cringles for lacing, or a jumbo size zip fastener in a curve just above the boom as shown in Fig. 46.

This removes fullness in neat and effective fashion.

The foot and leach of a foresail are usually controlled by the pull of the sheet, although leachlines can be fitted.

The leachline fitted to most mainsails can be tautened to increase fullness in the after parts of the sail, bringing camber aft. If the line is fully checked out, the sail will be as flat as its cut allows, but can be further flattened by bringing the boom down.

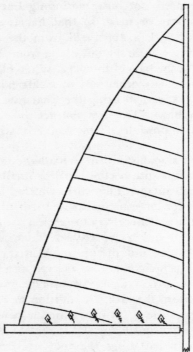

46. Flattening reef points

WINCHES

If you head at five knots into a Force 4 breeze, air pressure on the main will be 2 lb./sq. ft. or more, and that on the jib between 50 per cent and 100 per cent greater. To sheet the latter fully home would call for putting a couple of hundred pounds' effort into a sail of only 75 sq. ft., and this is beyond the power of the human arm. Some form of mechanical advantage has to be employed. This means using winches, plain or geared down, on all but minuscule craft. The mechanical advantage of an ungeared winch is found, roughly, by dividing the length if its handle by the radius of the barrel. It can be increased by using a longer handle which should, naturally, not foul its surroundings.

Luffs can also be tensioned with winches and a self-stowing model is almost essential for wire halyards; an open-ended one used in conjunction with a cleat is handy and tidy with rope. On biggish craft it may be necessary to use a geared winch to bowse down a large foresail luff.

Winches and other forms of purchase repay study, and the subject of this tackle and gear is dealt with in more detail in my *Handbook of Small Boat Cruising*, to which your attention is invited.

Mainsails are usually sheeted through a system of blocks but there are distinct advantages in taking the end of the sheet to a strategically placed winch for use on either tack. This cuts down the number of blocks and leaves less length of sheet lying around the cockpit when closehauled.

The end block of a mainsheet should be movable and a foresail sheet will lead to different points on deck according to size of sail and condition of sailing. An unfair lead to a winch will sooner or later cause over-riding turns; in strong winds these may jamb immovably and the only remedy is to cut through the sheet, which can be hazardous. The ideal lead depends on the individual winch, but in most cases should be angled up to the barrel at 5 degrees or so. To ensure a fair lead at all times, the sheet is best finally rendered through a suitably positioned static swivelblock.

SHEETING

The primary function of a sheet is to move the sail

laterally in order to set the angle α needed at the time. It also pulls the sail in a generally downward direction and that influences its shape. You should bear in mind that sails, and therefore the way in which they are sheeted, have to be looked at from the aspect of interaction. It will seldom be possible to have all sails sheeted with maximum efficiency; to get the best performance you have to arrive at the best compromise in sheeting.

Dependent on the condition of sailing (Chapter 4), one sail will be predominant in producing drive. It should be sheeted as effectively as possible and you will have to accept an inevitable diminution of efficiency in other sails.

SAIL TWIST AND WIND GRADIENT

If you ever see a square rig training ship under sail (or watch the 'Onedin Line' on the box) you may notice that sails are progressively sheeted more outboard higher up the mast. This setting takes advantage of the wind gradient, or fact that true wind speed increases with height because of surface friction having a retarding effect low down. An increase in wind speed for a constant boat speed will result in an increase in the angle (β) of apparent wind (see Appendix, p. 192). Your mast will probably not be over 35 feet in height above sea level; the difference in angle β will not, even in strong winds, be more than about 3 degrees between truck and step. If the head of a sail sets this amount more outboard than the foot, efficiency will be the same, but once the angle is exceeded, the upper and lower parts of the sail will produce different amounts of drive per square foot.

Fig. 47 shows a boat on a close reach. The mainsail can only be controlled by the sheet attached to the boom, which will angle high and let the head of the sail twist

47. Sail twist

badly away from the wind. To stop flapping and lost drive, the helmsman has to haul down on the boom; this also brings it more inboard. Only one part of the sail is then set at the proper angle α, the area above will be spilling wind and that below suffering turbulence and partial stall on the lee side. The best a helmsman can do is to balance losses by having optimum angle α about one-third the way up the sail so that areas above and below it are roughly equal.

The remedy is to fit a kicking strap so that the boom swings level laterally however the sail is sheeted. This helps to keep camber fairly constant over the sail. The kicking strap should pivot at a spot at the base of the mast directly below the gooseneck, otherwise the boom will not swing level. Also, its bitter end should lead from this point, and not from the boom eye, aft to the cockpit so that adjustment can be made easily when the boom is swung out.

Even if a kicking strap is used, the pull of a single-ended mainsheet or the resultant pull of a double-tailed one will be towards the midship mounting of the gear; this will tend to draw the boom too far inboard. The cure is to mount the sheet block on a horse, or track, extending as far outboard as feasible on both sides of the centreline. When the block is properly stationed you can heave down on the sheet without applying any midward-acting component of your pull. In close-winded conditions you can really flatten the sail even though the boom is well outboard.

If your boat is not fitted with a means of moving the sheet block, during trials you can temporarily secure it away from the midline by securing a short rope tail to the end of the boom and lashing it down to the rail. Remember to cast it off when going about!

If your boom is rigid, it does not matter from which part of the boom the mainsheet leads. There are many advantages in having the sheet forward of the cockpit, leading to a track on bridgedeck or top of coachroof. It keeps the working area clear of a flailing sheet when gybing or going about. Bear in mind that the further forward on the boom the sheet is attached, the greater will be the purchase needed to sheet in.

When a boom is flexible, and some of small diameter are quite bendy, it matters very much where the sheet leads from the boom.

With the sheet at the end of the boom, weight in the sail will stretch the leach while the belly of the sail increases (Fig. 48 A). Taken from the middle of the boom, the belly of the sail is flattened but the leach slackens off (Fig. 48 B). These results have different effects on camber and airflow, as you will by now be able to appreciate. The use of flexible spars is worthy of a chapter to itself, but as they are not really suitable for cruising boats I will

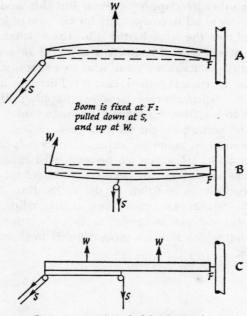

*Boom is fixed at F:
pulled down at S,
and up at W.*

*Fig. C shows method of minimising bend
by spreading both S and W loadings*

48. Boom flexion

leave the matter to those writing about racing.

Decades ago I used to sail a Dutch-built ketch of about 40-foot L O A which had a foresail set, loosefooted, on its own boom. Once stretched for the prevailing wind it could be forgotten when tacking or gybing, and to sheet it in or out was a simple matter of handling a single sheet taken to the cockpit. There was no kicking strap, I don't think they were used in those days, but today I would fit one as for the main boom.

Overlap was obviously impossible and perhaps I could have squeezed another half-knot out of *Josta* by using a genoa or other overlapping foresail but this would have gone no way at all to compensate for the ease of handling conferred by the club-footed jib. Crew comfort and absence of stress is as important as speed to a cruising crew; more, perhaps, to those who spend periods at sea and need to conserve their energy. There is an awful amount of frustration in handling flapping jib sheets attached to a sail which scrapes across the mast, chafing madly and getting caught on deck gear.

There must be room for experiment in this field. For instance, could a boomed jib be used with roller-reefing gear? One would need to devise a clew-haul taken back to the cockpit in addition to the roller line, but this should be within the competence of any sailor. I leave the thought with you and revert to the conventional, modern foresail which is much more difficult to sheet than is the main.

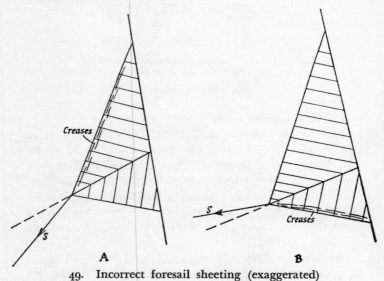

49. Incorrect foresail sheeting (exaggerated)

You can see (Fig. 49A) that a sheet angled too far down will pull harder on luff than foot and in 49B how the converse applies. With sheet pulling at the happy mean, the sail will adopt the shape to which it was cut and, when fully sheeted in on the wind, will set with optimum camber. (See page 48 for indicated amounts of camber.)

Once the sheet is slackened, no matter how little, the wind pressure will cause camber to increase from the foot upwards and also make the sail twist. Efficiency immediately drops. As you bear away from the wind, the increase in amount of the camber may get so great that the foresail starts to seriously backwind the main.

To a degree this difficulty is ameliorated by the fact that different windstrengths call for different amounts of camber as cut into the sail. Rising winds usually mean a reduction in foresail area, and the smaller sails can be cut with less camber. What has to concern you is the fact that the angle of trim (the angle between the chord of the sail and the centreline of the boat—δ_f for the foresail) is different for maximum efficiency in different wind strengths. In strong winds (except with a reefed mainsail, see page 188), the sheet should lead down to the deck considerably more inboard than in lighter weather. If the fairlead track is parallel to the C L this is not possible.

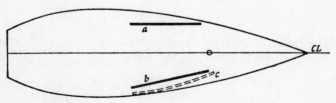

a—track parallel to CL
b—track angled out
c—curved track

50.   Sheet lead control

The track can be angled or curved so that the lead for a smaller sail used in strong winds will be more inboard due to being further forward. This only partially resolves the problem because in light airs you will want to use a large sail and also to increase its camber by taking its sheet lead forward. The requirements are mutually opposed.

A viable solution is to adopt a variation of the 'Park Avenue' boom idea and have two parallel sets of track on which a third one travels as shown:

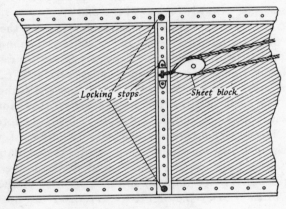

Locking stops   *Sheet block*

51. Travelling track

The fairlead is on the travelling track and can be located anywhere within the shaded area.

SAIL TRIM AND INTERACTION

Having installed what means of sail control you think necessary you can start sailing trials in earnest.

CONDITION 1

On the wind a foresail will produce more drive per square foot, up to 100 per cent in some cases, and should be given precedence in trimming. Angle δ_f will be smaller for stronger winds, but its size for any wind will depend on the characteristics of your boat. You will have to use speedometer and clinometer to indicate the effect of moving your sheet leads about until you arrive at the best setting for the wind of the moment. If your boat will sail with the mainsail down the matter is simple, although the best lead found may later have to be adjusted in the light of the slot effect. If both sails have to be used, trim the mainsail as best you can to match the wind, which will probably mean strapping it in hard in a strong breeze and centrally in a lighter one, and ignore it for the time being until you have found best angle δ_f. Then look to see if the main is being backwinded.

There can be a number of reasons for this. If the fore-sail leach is curved inboard, its slipstream will impinge on the lee of the main. That may be due to the sheet being too far forward, putting excessive tension on the leach. A 'hard leach', caused by the tabling being too taut, will have the same effect and can be remedied by easing the tabling. The sail can be too fully cambered so that it backwinds the main even though sheeting is correctly led. Generally speaking, the best combination is to have a flat foresail and a more greatly cambered main for windward work. Intermittent backwinding can be caused through 'motorboating', another matter for adjustment by attention to tabling.

Importantly, forestay sag puts extra camber into a foresail as the wind rises, which is just when it is not needed. It can be remedied either by tautening the stay, which should not be overdone, or recutting the sail with a hollow in the upper part of the luff, leach or both. You should not cut a perfectly good moderate weather foresail to this end without much thought; it may be better to consider buying a flatter one for hard conditions.

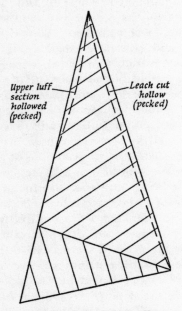

Upper luff section hollowed (pecked)

Leach cut hollow (pecked)

52. Reduction of foresail sag

It is possible to have a well-cut, accurately sheeted foresail and still suffer backwinding, in which case you have to look to the mainsail and, again, there are various courses of action open to you to rectify matters. The simplest one is, of course, to ignore it as long as you are getting enough drive from the unwinded parts which is often the case. In point of fact, many mainsails are too large in themselves for strong, closewind conditions, and should be reefed down so that a fairly large foresail can be carried as long as possible. If you do this, take care that lee helm does not intrude because you have taken the combined C E too far forward. (See Chapters 3 and 15.)

Full-length battening will assist a sail to hold its shape despite backwinding, and so provide considerably more drive. It is unfortunate that adherence to racing rules has stultified sensible and rational development of battening, which is an aspect of design particularly affecting cruisers. It seems to me that the matter could also be looked at in connection with foresails to prevent leach curl off the wind.

The camber of the mainsail may be excessive, which is perfectly all right in light airs but sticks its belly into the foresail slipstream in stiffer ones. Fig. 46 provides one answer to the difficulty.

Use of a horse, or track, will allow the mainsheet block to be taken out to weather and so bring the mainsail out of the slipstream.

By dint of experiment, as described, you will eventually discover the best sheeting compromise, and amount of sail fore and aft, which provides maximum drive and slot effect. It is a long business but worth pursuing.

If you are equipped with two headsails, the foresail should be trimmed for best effect; it is invariably larger than the jib which will probably have to be trimmed at a larger angle than the fore to avoid backwinding it.

CONDITION 2

On a broad reach the slot effect is much less than with the wind forward, so that you should be able to sheet each sail separately for maximum drive without there being any detrimental interaction. There is no problem with a boomed mainsail whose luff, leach and foot adjusters can be used to increase drive in all strengths of wind.

The foresail will begin to get inefficient because the sheet lead cannot be taken outboard to the point where it is fully effective in controlling camber and leach. There is not much that can be done about this, but you must never be tempted to sheet in too hard because the head of the sail is twisted, and perhaps flapping a bit. It is infinitely better to accept some loss of drive high up for the sake of much more lower down. Here, a hollow in the upper leach may prove advantageous.

As the wind gets further aft there will be blanketing by the mainsail. This you can partially combat by using a bearing-out spar to spread the foresail. It is possible that full-length battening of that sail would be beneficial, but this would rule out overlap.

Nearing running conditions, the essential thing is to spread as much sail as is warranted in the prevailing wind. The field is wide, with mizzen staysails, watersails, extra jibs and so on; everyone has his own pet ideas.

8

Auxiliary Power

The ubiquitous outboard is very handy as a means of harbour and inshore propulsion. As it can be unshipped or swivelled out of the water it causes no drag. Apart from putting weight right aft it has no detrimental effect on sailing performance.

The installation of inboard auxiliary power will almost always reduce sailing performance to some degree, due mainly to the propeller and its mountings causing drag and turbulence. The use of an engine can improve cruising ability and is in many ways a crew convenience, so that you have to compromise between them and pure sailing ability. Performance is in no way affected by such matters as choice between air and water for cooling, petrol and diesel for fuel, or bronze and stainless steel for shafting. It is also self-evident that bad maintenance and improper operation will make an engine inefficient. These things are left to your commonsense and not discussed further.

CHOOSING AN AUXILIARY SYSTEM

A few sailing boats are found here and there without any mechanical means of propulsion, but this may be taking purism to its limits of reason. It is virtually impossible to get in and out of crowded harbours under sail alone without causing frustration and possible damage to others. A small motor is a necessity for manoeuvring in confined waters and useful for covering short distances back to moorings when you are becalmed and the office beckons on the morrow.

Few boats are without the electric lighting and electronic instruments that add to comfort and safety, so that it is convenient to have a means of charging batteries aboard. An inboard motor can provide the facility, but few outboards are fitted with a generator. They are also not suitable for offshore use and it is preferable to instal an inboard engine.

The power needed to propel a clean hull of up to 30 feet overall at low speeds, say 3 knots, in harbour or calm conditions is surprisingly small; a 3 h.p. motor will suffice for such pottering. However, before settling on such a minuscule auxiliary, you will do well to query if these are the only circumstances you envisage for using it in the future. Having read what follows, it should be clear that a choice of installation lies between the very low-power auxiliary to which you should put an upper limit, and something of a much higher order of output to which you should put a lower limit. Anything in between these limits will be unsatisfactory, like anything else that is just too big or just too small.

In this connection, people often advance the question of fuel consumption. If you habitually use motor as an alternative to sail, which is a perfectly reasonable thing to do, it is best to regard your craft primarily as a motor-boat and equip her accordingly. The question of fuel consumption is then important and your choice of installation will need very careful consideration. If the engine is used only when it is not possible to sail, the probability is that the number of hours run in the course of a season will be so small that specific consumption is irrelevant on grounds of cost. On the same score, the saving in fuel costs incurred by fitting diesel in preference to petrol engine may never overtake the extra cost of the heavier and more expensive item. There is a lot of sloppy thinking in this field.

Unless you are lucky enough to have unlimited leisure time, your arrivals and departures on passage will have to be closely regulated. It is usual to plan a cruise on the basis of an average speed and inexperienced skippers are prone to overestimate it. As a rough guide, a rational average can be estimated by applying a speed/length ratio (R) to your boat of about 0.90 to start out with; it can be modified in the light of experience. $R = 0.90$ works out at a boat speed of 3.8 knots for a L W L of 18 feet, 4.0 knots for 20 feet and 4.2 knots for 22 feet. These figures are realistic for cruising although, of course, you may make much higher speed over short passages.

To maintain the average speed you will have to sail at speeds approaching V_{max} for much of the time. To make up for lost time caused through lulls and calms, you should be able to motor along at a speed approaching V_{max}. For this you will need considerably more power than is needed for pottering, and it is not easy or straightforward to estimate your requirements.

To begin with, the term 'horsepower' is very loosely used, as well exemplified by my own inboard and tender motors. One is a 4-cylinder, 4-stroke of 998cc capacity and described as a '10 h.p. marine engine'. The other is a single cylinder, 2-stroke of 49cc and termed the '3 h.p. model'. This does not seem to make much sense, unless you realise that there is no common standard of comparison.

The figure for the outboard is 'nominal', or theoretical, horsepower. That for the inboard is 'brake' horsepower, the criterion usually used for marine engines; it is the power arriving at the point of propeller shaft coupling, after the nominal h.p. has been reduced on account of transmission losses, power to drive the generator, and other sources of loss. The horsepower given will be the power at full throttle with the engine turning at a stipulated number of revolutions. If the throttle cannot be fully opened nor the engine turn at the proper speed, full power will not be obtained; this must be considered in choosing a propeller. B.h.p. is further reduced through friction so that the power arriving at the propeller for conversion to forward thrust (delivered h.p.) is some 15 per cent less than the b.h.p. quoted.

It is undesirable to run at full throttle for lengthy periods, and manufacturers also quote 'continuous output rating'. This is the power produced at an engine speed which can be maintained indefinitely without mechanical penalty. You should match your propeller to this speed, as will be explained; it will still serve well if you have occasion to use full throttle for any reason.

Power requirements rise rapidly once you exceed a speed equivalent to $R = 1.1$, so that to reach your V_{max} of perhaps $R = 1.275$ you would need twice the power needed at the lower value. If you can decide to hold your cruising speed down your engine can be of much smaller output than if you want to travel at maximum attainable hull speed. The following table is representative of the requirements of a moderate displacement motor-sailer of 25-foot L W L:

Speed (knots)	b.h.p. required	$R =$
4.0	4	0.80
5.25	8	1.05
5.75	10	1.15
6.25	15	1.25
6.50	20	1.30
6.70	?	1.34

This shows clearly how much extra power is needed to attain higher speeds until, when R reaches its maximum of 1.34, very little more speed can be attained no matter how much power is used.

Such figures are, however, somewhat idealistic, and assume that an engine will always produce its rated power; that it will be accurately matched to a fully efficient propeller; and that the boat being driven has a hull free from fouling and other sorts of drag. None of these conditions is likely to be met with simultaneously in real-life conditions. An engine can be out of tune or only partially effective. A propeller can be out of true, bent, defective, fouled or improperly matched to output. In some conditions a hull can foul up badly with marine growth during a week's cruise.

Attainable speeds quoted are for still conditions and you can easily envisage others where more power might be urgently needed. Headed into a 30-knot wind, head seas and a 4-knot stream bearing you into hazard would be an unappealing situation if your sails were damaged or you had no room to beat out. To just hold your ground you would need to move through the water at 4 knots; the power needed for this would then have to be augmented to overcome the windforce, which alone could force you at 3 or 4 knots astern, and the added drag caused through the boat's motion in the seaway. To move out of your position would call for yet more. The total power needed might well be twice that required to travel at cruising speed in calm conditions. It is wise to allow reserves of engine power for emergency use; how much this should be is a matter for judgment.

It is, of course, not only in emergency that you may need extra power. Many folk prefer to use their engine rather than make to windward in a strong breeze and lumpy sea. The extra resistances mentioned above will also then be present, and you could need as much as 50 per cent more power than for still conditions even to achieve a speed reaching R=0.8. This needs to be thought about in connection with maintaining a cruising schedule.

The weight of an engine is almost irrelevant to performance under either sail or power, and quite often a powerful one will weigh less than one of only half the output. Its weight makes little difference to the displacement weight of a cruising boat (see t.p.i. factor on page 200) and it is largely self-limiting. You cannot fit a disproportionately bulky and heavy engine into a small hull; there is insufficient space available.

PROPELLERS

A large slow-turning propeller is more efficient at converting torque into thrust than is a smaller one turning more rapidly. Ideally, you should fit the largest propeller that your hull will accept, but sailing performance has to be thought about. Propellers and their mountings create drag, and there may be a limit to the amount you are prepared to tolerate. A racing skipper might wish to keep drag to a minimum and put up with consequential inefficiency under motor. The owner of a motor-sailer will probably go for good performance under power and shrug off the fact that sailing ability will be restricted. It boils down to compromising between thrust and drag.

EFFICIENCY

The ability of a propeller to convert the power delivered to it into boat speed depends on its diameter, pitch, blade design, number of blades, and its speed of rotation. Any propeller will be most efficient at a set speed of rotation but will, obviously, be effective at all engine speeds. The engine or propeller-makers will advise on the ideal propeller to fit to your boat, in the knowledge of engine and hull characteristics. Other propellers within a range of efficiencies can be fitted if circumstances preclude use of the ideal. For instance, if you were recommended to use a 14-in. (diameter) × 12-in. (pitch) screw, you could also get reasonable results from one measuring 16 × 10, 15 × 11, 13 × 13, or 12 × 14 inches. The rule of thumb is to increase pitch by the amount by which diameter has been reduced, or conversely. The adjustment should not be more than about 15 per cent either way because efficiency is greatly influenced by pitch ratio i.e. $\dfrac{\text{pitch}}{\text{diameter}}$; by excessively changing the relationship between pitch and diameter, you might alter pitch ratio to an unwarrantable degree.

The recommended propeller will always be the largest that will fit your hull, and as it is usual for large propellers to turn relatively slowly, you may have to drive it through a reduction gearbox of between 2 : 1 and 3 : 1 ratio. Such a screw gives maximum efficiency for the power available. It does not matter theoretically whether it has two or three blades, provided that they are wide, and in some circumstances a two-bladed screw will be more efficient, although marginally, than one with three similar blades.

The results to be expected from a propeller will be affected by its location. One sticking out to the quarter, where it is in a region of relatively untroubled wake, will be capable of producing more thrust than another which is partly shielded abaft the central deadwood. If fitted to a quarter bracket your propeller will be slightly angled to the fore and aft line; this has little effect on its efficiency as such, but it should be sited with regard to its 'hand', or direction in which it turns.

Looking at a propeller from astern, one which turns clockwise is 'right-handed' and a 'left-handed' one will turn anti-clockwise. A right-handed screw will tend to turn the stern out to starboard and conversely. This effect can be countered by fitting it on the starboard quarter, where its angle with the C L will give it a tendency to shove the stern out to port: a left-handed screw should be fitted to port.

If set behind deadwood, or skeg, a two-bladed propeller will enter and leave the waterflow during each revolution of the shaft. This not only cuts down its thrust, but causes vibration which can be very objectionable. A three-bladed screw in the same position will always have part of at least two blades actually in the wakestream; more drive will be available and vibration will be minimal or absent.

DRAG

A large propeller obviously offers more drag than a smaller one, so that if drag reduction is to take precedence over thrust production you will wish to fit as small a screw as is practicable. Also to locate it to offer minimum resistance to the waterflow. In practice there is a lower limit of dimensions which cannot be exceeded if the propeller is to continue to absorb the rated engine output decided on.

Reduction in size means an increase in shaft revolutions. It is not normal to use step-up gearing, and propellers within the small range will nearly always be driven at crankshaft speed.

Low-drag installation means siting the screw behind the deadwood, or skeg, and if possible shielding it so that when sailing it is not in the waterflow. This calls for the use of a narrow two-bladed propeller which can be locked in the vertical position. If the deadwood is wide and cut

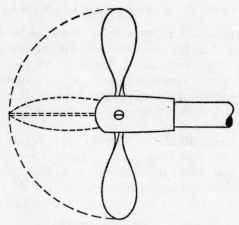

A *Clamshell Propeller*
When at rest, blades fold (as dotted).
As shaft rotates, they open auto-
matically under centrifugal force.

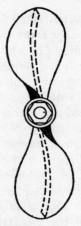

B *Feathering Propeller*
Blades rotate to offer least resis-
ance to waterflow (dotted). This
must be done through linkage and
is not automatic as with the
folding type.

53. Adjustable propellers

off square, it will most probably cause turbulence and more drag than if it were streamlined and the propeller projected slightly on either side of it. To lock the propeller, it is only necessary to set it up and down behind the deadwood and then put the engine in gear; the shaft can be marked to indicate the position of the propeller with a spot of paint.

Other means of minimising drag are to use a feathering propeller or a folding blade (clamshell) type. These produce less thrust than models with fixed blades, but with the clamshell blades folded inwards or the feathered blades set parallel with the waterflow drag is greatly reduced.

Such designs are useful if there is no deadwood and propeller and shaft have to be fitted in a bracket under the aft bilge. The bracket should be streamlined as far as possible.

A question that is often asked is whether it is better to let an exposed propeller rotate when the boat is being sailed, or to lock it solid. The answer is that it depends entirely on how freely it can rotate. One which can spin fast will cause less drag than one which is locked. One which is impeded in rotation by transmission friction in gearbox, stuffing box, and so on may cause more drag than either a locked or freely rotating one. There is no simple way of finding out except to sail and see what effect there is on boat speed by firstly locking the screw and then releasing it. A propeller offering great drag, like a three-blader out on the quarter, could be fitted with a sailing clutch. This frees the propeller and its shaft completely from the transmission, and it rotates very freely. Some yacht propellers are designed with a profile enabling them to revolve faster than would one with the conventional, high-thrust profiling. One of these used in conjunction with a sailing clutch would probably provide the best answer to the problem, but if you sail a lot there is constant wear and perhaps irritating noise as well.

CRUISING PERFORMANCE

Cruising performance can be enhanced by a judicious use of auxiliary power when under sail. A light breeze aft of the beam can be converted into a stronger one forward of it, providing a reaching attitude, if progress is gingered up by a knot or so from the engine. Use of power is also helpful in making more to windward if your craft is not particularly weatherly.

Such expedients are often sneered at, but I see nothing wrong in using whatever means are available in order to cruise more successfully. The odd couple of hours under power-assisted sail will charge batteries, exercise the motor and perhaps make the difference between missing a tide with a six-hour wait or a tedious flog against it with motor roaring flat out, and a timeous arrival while the shops are still open.

9

Drag: the Arch-Enemy

As a boat exists in the interface of water and air, her abilities are governed by the physical laws applicable to both media, which are fluids differing only in degree. Performance will depend on the way in which you gird your boat both to take advantage of the beneficial effects of these laws and combat the disadvantageous ones. Up to now you have been mainly concerned with increasing propulsive power, but it is equally important to neutralise drag to the greatest possible degree. Drag has been given passing mention in previous chapters but you may find it of benefit to consider the matter a little more fully.

INDUCED DRAG

Sails of high A R are of advantage on the wind because of the extra drive afforded by their length of luff, but this also helps to keep down induced drag which arises from eddy, or vortex, effects.

Air will always try to move from a region of high to one of lower pressure, as previously explained. It will always manage to curl around both head and foot of a sail and this results in trailing vortices, as shown in Fig. 43. The effect is greater with close winds and falls away to nothing as it broadens off.

In the case of a high A R sail, proportionately less air can get from one side to the other and vortices will be smaller and cause less induced drag. The influence of the foresail, especially if taken to masthead, reduces the length of leach of the mainsail head over which curling takes place. The length of the foot clearly limits the area of action in that region; the endplate effect of a wide boom and the shape of a sail in ameliorating induced drag were mentioned previously.

As induced drag is also of consequence hydrodynamically it is worth examining the problem in a bit more detail. A flat plate lying parallel to a flow of fluid along it will have no pressure differential about it—see Fig. 54. When it is inclined at a small angle to the flow (Fig. 54A), pressure rises on the front side and drops on the back due to differences in flow velocities, thus causing the pressure differential dP. If this angle of attack (angle α) increases, a stagnation point appears near the leading edge of the plate (S) where the fluid has ceased to move and simply exerts a positive pressure (54B). As the angle gets still larger, the point enlarges into an area of stagnation and flow over the back breaks away from the surfaces and causes eddies (vortices). This is, of course, the angle and point of stall where dP starts to get smaller (54C). A stage then ensues when eddies form at both leading and trailing edges, and finally there is a situation when eddies of equal size and effect are found when the plate is at right angles to the flow. This is the state where a sail directly before the wind gives rise to rhythmic rolling as the eddies alternately break away from the lee side and travel downwind (54D).

Sail stalling causes a loss in driving potential, but underwater eddies about a hull offer resistance to forward motion. If parts of the hull are blunt-ended such as square-cut keels, rudders, deadwood and so forth they

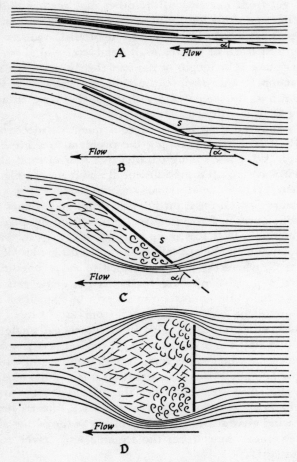

54. Flat plate in waterflow

will absorb considerable power. The resistance of a plate at rightangles to a flow is composed of about 75 per cent head resistance and 25 per cent due to induced drag caused by eddying astern of it. The same 25 per cent will exist even if the head resistance is done away with by streamlining the forward side of the plate, as shown:

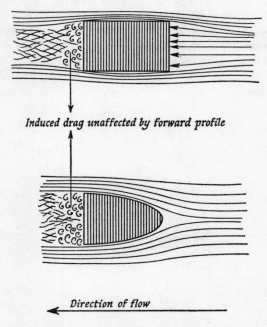

Induced drag unaffected by forward profile

Direction of flow

55. Induced drag in wake of object in waterflow

It can also be almost eliminated by streamlining aft, so that blunt entries like square stems and blunt ends as described can both be faired off to present minimal resistance to the waterflow. This is effective because the waterflow is generally from fore to aft along a hull, but the airflow can be from any direction making it impossible to give streamlining to upperworks and rigging which will invariably be useful.

For instance, a mast which is designed to present minimum resistance to a beam wind would offer maximum area to one from astern. This would, in fact, be of advantage in increasing the area offered on a run, but would be very disadvantageous when beating. With the exception of crosstrees and jumper struts, which can be streamlined in section from fore to aft, the best compromise for mast, spars and rigging is to fit circular section. This will offer the same resistance irrespective of wind direction.

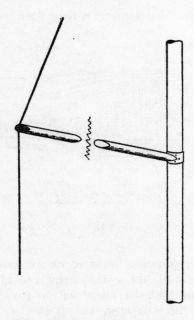

56. Streamlined crosstrees

FRICTION DRAG

Fig. 57 shows a curved surface in a flow of fluid which can represent either a sail in the wind or one side of a hull under water.

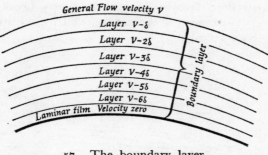

57. The boundary layer

Neither air nor water can flow over a surface without friction which causes particles to stick and remain stationary with respect to it. The next layer of particles outward will move slightly in the flow and the effect spreads until there is a final layer moving at flow speed. The region between the stationary and outer layers is known as the 'boundary layer' of the medium. It is inconstant, and the way in which it changes is critical from the viewpoint of drag. It behaves in accordance with a 'Reynold's Number' which is determined by four factors, viz., rate of flow, length and smoothness of surface acted on, and viscosity of the fluid. For a given viscosity, in our case either air or water, the R N will vary directly according to length and smoothness of surface and rate of flow.

At low values of the R N the boundary layer will stay parallel to the surface, a condition known as 'laminar flow', when friction drag is virtually absent. At a certain point where the R N rises above a limit the flow becomes altered; the layers of particles become agitated and start to oscillate, which causes the boundary layer to get wider and to begin to offer friction drag to the flow. This effect can be seen in the white and bubbling turbulent water running along the hull and widening as it goes aft—turbulent flow. Eventually, this flow may break down and eddies form next to the surface, which will cause drag to rise considerably.

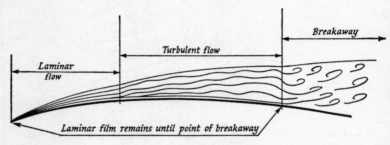

58. Analysis of flow

The points at which turbulence and then breakaway are initiated depend on the length of the surface, its smoothness and its profile, or shape of curve. Laminar flow persists later on smooth surfaces than on rougher ones which accounts for the frantic polishing of bilges by racing dinghy fanatics; it really pays off because the friction drag of laminar flow is so low in comparison with the other types that it is worth making it as lengthy as possible over a hull.

Any prominence or projection on a surface will disturb the flow and its effect will spread out fan-wise (Fig. 59). It is thus important to avoid excrescences in regions where flow might be laminar, as along the foreparts of a hull. Items like impellers, transducer heads, skin fittings

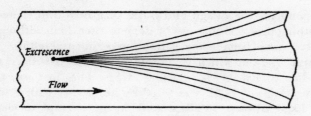

59. Spread-out of turbulence from small protuberance

and so on will give increasingly less drag as they are fitted further aft. Flow tends to be more prone to turbulate and break away once it has passed the point of maximum curve; it therefore does less harm to put your fittings aft of the section of maximum beam (on the waterline) than forward of it where they may give rise to disturbance in a region which is not naturally prone to initiate turbulence and breakaway.

Similar considerations apply to sails. Due to the last of these tendencies, the point of maximum camber is important. Unless a slot effect is used to keep the airflow close to and flowing fast along the lee of the mainsail, it pays to have camber as far aft as possible without risking the introduction of an astern-acting component of thrust.

For equally cambered and smoothed sails, the point of turbulence and then of breakaway will be at the same actual distance aft of the leading edge. A low A R sail, with longer chord, will produce more drag for a given area than one with higher A R. Added to the lesser pressure differential afforded by the shorter luff and greater induced drag, this fact accounts for the relative inefficiency of low A R sails on the wind.

There are other sources of drag than those mentioned above, such as wave-making resistance and the induced drag of underwater parts of the hull. In general, these depend on the design and build of a boat and there is nothing practical that can be done to alter them, although there have been some gallant experimental efforts. A blunt transom which is constantly immersed will create eddies astern of it, causing drag. The first of the Macwesters, *Sabepe*, was fitted by her owner with a sort of fixed trimtab, as illustrated, in an attempt to clean up her wake (see frontispiece).

10

Trim

One of the most common and least recognised reasons for consistent under-performance is that a boat does not sail on an efficient waterplane at all times. Really poor trim will significantly affect behaviour and you should take a critical look at your own boat from this aspect.

As stated in Chapter 1, a boat is designed around a set of lines and these can be changed in shape without physical alteration to the hull. When a boat rolls or pitches her waterlines, and the waterplane they bound, become distorted as do underwater sections and other shapes. Fig. 9 has shown how heeling alters the waterline shape. This effect is irremediable, but good designers take much trouble to minimise the undesirable effects over as large a range as possible. However, waterlines can be distorted for other reasons, the cure for which lies in your own hands.

If a boat trims down by head or stern, it is apparent that her waterlines will then not be as designed irrespective of whether she is heeled or upright. To confirm this for your own satisfaction, draw in a fresh W L on the sheer plan at a small angle to the D W L, say 5 degrees; the lines should cut at a point vertically above the Centre of Buoyancy (C B)—see Appendix p. 198. With your dividers, transfer the measurements to the half-breadth plan and you will be able to draw in the waterline corresponding to the angle of trim concerned:

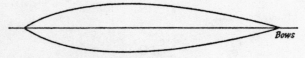

LWL of boat trimmed down by the bows

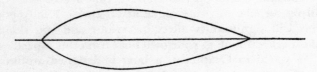

LWL of same boat trimmed down by the stern

60. Changes to L W L

It should be clear that if you start off with an unfair waterplane of this sort, the effect of any motion such as heeling or pitching can only make matters worse.

The root cause of a boat being out of trim is that her C G lies forward or aft of the designed position. This will inevitably be due to bad stowage of cargo, including any inside ballast, and the placement of crew members who weigh about twelve to the ton.

When a vessel is floating freely the C B and C G are vertically in line. If the C G shifts, the C B will re-align itself accordingly. By definition the C B is the *centre* of buoyancy and the amounts of water displaced fore and aft of it by the hull must be equal. If the C B is shifted, as when re-aligning itself with the C G, the volumes of water fore and aft have to redistribute themselves and the end result is a change of trim. Once again, you might care to examine the effect of moving the C G of your own boat 6 inches aft of its designed position.

Measure the areas of the underwater sections at the old and the new C G positions, add them together and divide the answer by two. Multiply the result by the distance between sections to give the included volume and then divide that by 35 to give the weight of water in tons:

$$\text{Area} \frac{\text{(old)} + \text{area (new)}}{2} \text{ sq. ft.} \times \tfrac{1}{2} \text{ ft.} = \text{V cu. ft.} \qquad (1)$$

$$\frac{V}{35} = \text{W tons of water} \qquad (2)$$

This is not absolutely accurate because the volume of water is in reality wedge-shaped, but this can be ignored for purposes of illustration.

The weight W lay aft of the original C G/C B line and must now shift to be aft of the fresh one. That can only happen if the after part of the hull sinks a corresponding amount and there is an equivalent rise by the bows. This agrees entirely with what everyone knows without saying, that a weight in the after part of a boat will depress her stern, but it is essential to realise that it is the alteration to the position of the C G which matters.

The remedy is obvious. You should take pains to keep your boat level with her D W L. She will sail best when actually lying to the D W L because all lines are designed around it but due to different loads will usually be a little above or below it. The weights of crew, stores and dunnage in smaller classes represent a large percentage of all-up weight and must be stowed intelligently so that the boat lies on, or level with, her D W L as often as feasible.

Crew movement is intractable due to sailing demands, but inept stowage will seriously affect performance. This is an argument for not fitting large, built-in tanks, especially fuel tanks which for reasons of safety and convenience are usually sited right aft. Any such tanks should be kept filled to a uniform level and replenished as soon as possible after any great amount of fuel has been used. The same thinking applies to water tanks, although they are less objectionable if sited about the general area of the C G when variation in the weight of their contents will not matter anything like as much.

I dislike built-in tanks for more than one reason. They are seldom fully emptied and drained off, so that fuel tanks tend to accumulate sludge and water tanks to become fusty. Apart from this, liquids are quite heavy, a Jerrican full of water weighing 45 pounds and one of petrol about 35. They can be used as a form of inside ballast, a quantity of which is useful to have aboard for trimming purposes. This does not imply loading up with great lumps of lead, but arranging alternative stowage positions for weighty objects which are customarily carried on passage, like chain and anchors, batteries and so on. The positioning of inside ballast for best effect is dealt with at length later on.

EFFECT OF TRIM ON CLR

The position of the C L R changes with trim, going forward as the bows dip and conversely. You can illustrate this effect by cutting out and balancing the underwater lateral areas resulting from the insertion of inclined waterlines on the sheer plan. Refer back to Fig. 7. Better than this is to experience the effect of a moving C L R in a vividly practical way.

Most boats will sail themselves for a limited period when on the wind, if sails are carefully balanced against rudder and the helm either held or lashed immovable.

Get your boat sailing like this in smooth water so that seas do not knock her about. Send a man from cockpit to pulpit; your boat will alter course a few degrees to windward and then settle down on a new heading. As the man returns aft your boat will reverse the process and, hopefully, regain her original course. Course changes of this nature arise from the changed reaction between the wind acting at the C E of the sails and the resistance of the water acting at the C L R. This reaction is basic to a boat's behaviour under sail and you will need to keep it in the forefront of your mind throughout all sailing trials and, indeed, whenever paying attention to behaviour afloat.

SAILING FORCES

When a boat moves about in a seaway the C L R is never in the same spot for more than a moment. Increase in speed causes it to move forward; pitching is, in effect, an alternate trimming by head and stern and the effect of that on C L R has just been pointed out; finally, heeling brings about changes in the waterlines which also have an effect on the C L R.

Chapter 3 made it plain that the C E of the sails is also constantly moving for a number of reasons. The movements of the two separate loci are in no way interconnected and it is, in any case, quite out of the question to try to estimate their positions simultaneously and continuously. In allowing for the reaction between two moving centres like these, you can only take their designed positions as a basis for action, and plan adjustments to your boat and her handling accordingly. This will mean, fortunately, that in practical terms a *pro rata* adjustment will be reflected on to the actual loci, despite their transient whereabouts.

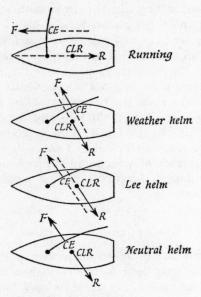

61. Forces acting on a boat

The drawing shows the principal forces acting on a boat under sail. Disregard the direction of the wind of the moment and accept that it acts in the direction shown along the line C E—F and also that the resistance of the water is acting similarly along C L R—R. Unless the vectors (lines showing both force and direction) are opposed to each other there will exist a rotating couple tending to turn the boat into or away from the wind. Due to movement of the loci the couple will vary in strength and, perhaps, direction of rotation. If a boat is 'neutrally balanced' so that there is no appreciable weather or lee helm felt at a certain angle of heel, it is possible that the direction of rotation will be in either direction: at times the boat will carry lee helm (usually

in light airs) and at others you will feel increasing
weather helm. This matter is thrashed out later under
the heading of 'Yaw'.

The extreme instance of reaction between wind force
and water resistance is when a boat is running, not goosewinged or under spinnaker. The C E is then so far out
to the side that there is no chance whatever of the forces
becoming diametrically opposed. This is why a boat going downhill needs large tiller movements and shows
a distressing tendency to broach. I have digressed a little
from the subject of trim, but it is important to hoist in
this relationship between forces as early as possible; it
crops up everywhere.

BALLAST

All weight carried aboard can be regarded as ballast in
addition to fixed ballast in the shape of fin or bilge keels.
In addition to its effect on trim, the proportioning and
placement of ballast affects roll (transverse stability) and
pitch (longitudinal stability), so that in making adjustments to weight you must take these effects into account.

11

Yaw

Yaw is the rotation of a hull about a vertical axis. It is caused by the effect of wind and water upon a hull and it is controlled by the rudder.

A hull floating in still conditions will yaw about the axis of lateral resistance. It will try to do so when sailing, but as explained above a rotating couple may exist, in which case the boat will rotate about an axis centred somewhere between C E and C L R. It is desirable for a boat to have a tendency to turn into wind for reasons of safety. Lee helm should not be tolerated except in very light airs, for an untended helm will allow a boat to bear away, gybe and perhaps broach. If the rotating couple turns her to weather when the helm is released, it is often known as a 'weathercocking' couple.

To overcome the weathercocking tendency you have to apply a bit of weather helm; this may increase as the boat heels more, but with a well designed hull it will seldom build up to an intolerable extent as long as sail area is sensibly matched to the wind strength. Indifferent design may cause a boat to be piggish, even unmanageable, at large angles of heel.

I would like to distinguish between *angle* and *weight* of weather helm. To a helmsman, 'weather helm' usually means the pressure on the tiller. This may be uncomfortable but still have no detrimental effect on performance, whereas a tiller which feels light and responsive may be slowing your boat down quite significantly. It is a question of the amount of drag produced by rudder action, and this is mainly a matter of angle and not the weight experienced at the helm.

The drag of a rudder is conditioned by its size and shape, very much in the same way as the drag of a sail, so that the first action you should take is to check on its area. The immersed part of the blade should measure between about one-tenth and one twelfth of the lateral area of the underwater part of the hull. This can be checked with your planimeter. A skeg assists rudder action, so that a rudder set behind one can be relatively smaller than other types.

Helm weight can be divided into the pressure needed to rotate the rudder, which is present whether the boat is under way or not, and pressure transmitted from rudder to tiller as a consequence of movement through the water.

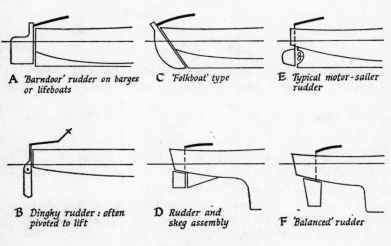

A *'Barndoor' rudder on barges or lifeboats*

C *'Folkboat' type*

E *Typical motor-sailer rudder*

B *Dinghy rudder: often pivoted to lift*

D *Rudder and skeg assembly*

F *'Balanced' rudder*

62. Types of rudder

Compare types A and B in Fig. 63. The effort needed to move a rudder through a given angle in still water is a function of its aspect ratio.

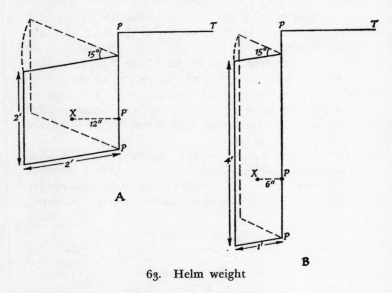

63. Helm weight

The areas are the same at 4 square feet and both rudders move through 15 degrees. The volume of water displaced by type A will be $\dfrac{15}{360} \times \pi \times 2^2 \times 2 = \dfrac{1\pi}{3}$ cubic feet; the volume displaced by type B will be $\dfrac{15}{360} \times \pi \times 1^2 \times 4 = \dfrac{1\pi}{6}$ cubic feet. This is half the quantity.

Further, resistance to movement will be centred at the Centre of Pressure (CP) of the blade (see Appendix, p. 204). With tillers of the same length the mechanical advantage for rudder A is length PT divided by length

XP i.e. $\dfrac{P\,T}{1}$ = P T. For rudder B it is $\dfrac{P\,T}{\frac{1}{2}}$ = 2 × P T. This is twice the leverage.

Twice the leverage to displace half the quantity means that rudder A will need 4 times the effort to move it through the same angle as rudder B. You may not have to move it through the same angle to produce the same change in heading, but the initial weight felt will be greater, which is why some skippers complain of 'heavy helm' or 'heavy rudder'. The effect is very noticeable at low speeds when other forces acting on the helm are low in comparison.

As I said, a rudder controls yaw; it exerts a force which can turn a boat away from its heading. In some conditions it cannot do this, and the boat then has 'no steerage way on'. It is quite normal to be in this state at very low speeds, and causes difficulties to inexperienced skippers in harbour conditions of mooring up and casting off. More baneful and infinitely more dangerous are conditions of no steerage way when under sail, perhaps at hull speeds. These arise with following seas, and are to be encountered in unpredictable water like minor races such as are found off St Alban's Head and Start Point. The lack of control is because the rate of water flow over the rudder is too low to produce an effective turning force; some improvement can be gained by an increase in blade area.

A rudder will be most inefficient if a section has been removed from its leading edge in order to accommodate a propeller as shown in Fig. 64.

This has a very detrimental effect on the ability of the blade to provide adequate turning force and if the missing section can be filled out improvement will be dramatic.

A rudder acts by turning in the water flow and producing a force F at roughly right angles to the blade

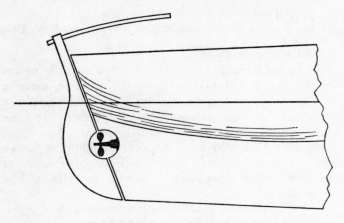

64. Rudder cut away to admit propeller

surface. This is broadly analogous to the force produced by the passage of air over a sail, or aerofoil, but has a much more powerful effect for a given area due to the greater density of water.

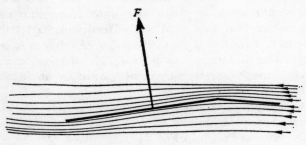

65. Rudder in waterflow

Force F is resolvable into a turning component F_t and a component of drag F_d which is resistance to forward motion. F_t acts on the lever between the C P of the blade and the axis about which the boat rotates, thereby setting up a yawing movement. If the boat is to sail a straight course, this moment must exactly balance out the weathercocking, or lee-turning, couple caused by wind pressure. If the couple is great the helm angle will have to be correspondingly large and so increase F_d at the expense of F_t.

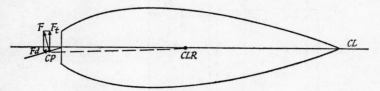

A *Turning force F_t and lever CP-CLR*

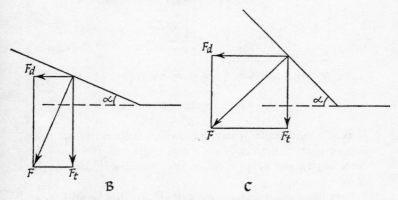

Increased drag with greater angle of rudder to waterflow

66. Rudder forces

Before dealing with the drag that is so inimical to progress I would like to dispose of weight on the helm, which is irksome and possibly tiring.

The pressure on an object in a waterflow is a function

of the *effective* area presented at right angles to the flow. Assuming speed to be constant, Force F will vary according to the angle at which the blade sets because this alters the effective area.

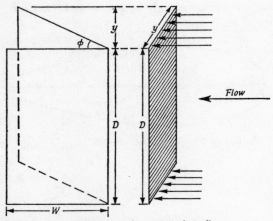

Effective area of plate turned at angle φ to flow
= y × D (y = W sin φ)

67. Effective area of rudder

It is at a minimum when the rudder is fore and aft and a maximum when it is at right angles to the flow. It increases in direct proportion to the sine of the angle of attack, α.

Pressure is exerted at the CP of the blade and, precisely as for the static pressure mentioned above, is felt at the helm according to the ratio XP : PT (Fig. 63). The weight felt with a blade of high AR will be proportionately less. Thus, one remedy for reducing weight of helm is to increase the AR of the rudder blade.

This applies to most types of rudder shown in Fig. 62, but type F is unique. Between one-sixth and one-ninth of its total area lies forward of the pivot line. This auto-matically brings the C P forward and reduces the pressure transmitted to the tiller. In extreme cases there is virtually no pressure felt and the helm is dead. Changes to the weight felt will depend in this case not on the size of blade or its A R, but on the proportions fore and aft of the pivot line.

Drag comes from a number of factors. Skin friction due to passage of water over the blade can be reduced by keeping it smooth and clean. A R is irrelevant here because the total area exposed to the flow can only depend on size, not shape. You have seen that increased angle α results in increased F_d, but that is not the whole story.

Depending on its section, which can vary from flat to streamlined, any rudder has a maximum angle of attack beyond which it starts to stall, with the formation of turbulence and eddies, exactly as does a sail and for the same reasons. The inevitable result is that the dP between the sides of the blade drops, and so does turning force F_t. F_d, on the other hand, is augmented by the induced drag of turbulence and eddies. In an effort to find more F_t a helmsman will increase rudder angle and this just makes matters worse. Eventually he reaches the stage of presenting the blade at such a large angle to the flow that the boat slows down to the point where the partially ineffectual rudder is capable of controlling it. It does this by running up into the wind to the point where drive is lost sufficiently to let the rotating couple diminish to such a size that the available F_t can balance it.

On the wind this merely causes loss of way to some degree, but if you happen to be running the situation is more fraught. The rotating couple is very large (Fig. 61). If the maximum tiller movement cannot provide enough force to prevent the boat running up into the wind, she will have to turn through almost 180 degrees to do so, and this is the classic broaching situation. The boat accelerates in a curve, the helm is hard over and the rudder completely stalled out. Under plain sail, the only danger is that of a beam sea, but if a spinnaker is carried it will remain filled when the boat is beam to the wind and a knockdown can result. A goosewinged foresail can also cause trouble by filling in a backed state.

It should be clear that you cannot expect good performance if the weathercocking couple is so large as to make you take your rudder to near, or perhaps beyond, the angle of stall under normal conditions. There is nothing left in reserve. If you want to bear away you will exceed the stalling angle and the boat slows down. In gusts, the couple will probably increase and need more helm to counter it; the result is always the same—loss of speed and perhaps control.

It is important to realise that the *weight* on the helm has nothing to do with the matter. As explained, a rudder as shown in Fig. 62F could be stalled right out without much pressure being felt on the tiller.

The leading edges of types 62B and 62F enter directly into the water flow and cause an abrupt change of direction as they are angled into it. This is conducive to turbulence and such rudders will stall abruptly and likely somewhat earlier than would 62A, C and E. Type 62D is intermediate as the skeg prevents leading edge meeting direct flow.

The sectional profile of hanging rudders, and of a rudder and skeg assembly, is important from the viewpoint of efficiency.

Force F acts at right angles to the flat plate A but is inclined a few degrees forward of this with the fully streamlined profile B. This means that F_d is less for the same angle ø, and F_t greater.

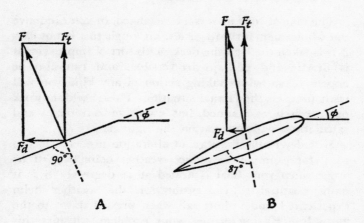

68. Effect of streamlining a rudderblade

The angle of stall is some 2 to 3 degrees greater with a streamlined profile, and such shaping gives advantages all round. Hanging rudders can be modified to a streamlined form by means of bodying them up with laminations of resinglass. Rudders behind skegs or deadwood can be faired off on the trailing edge to prevent the formation of eddies astern of them. If it is not feasible to streamline a rudder and skeg assembly fully, you should at least try to see that all leading edges are parabolic in profile and trailing ones not left in any way blunt.

To summarise, the most efficient type of rudder is probably one of high A R, fully streamlined, and set as close as possible to any skeg or deadwood to prevent pressure leakage and eddies on the leading edge (v. gaff sails). If you cannot change or modify your rudder to combat excess helm and its effects, you may have to tackle the problem somewhat differently. Instead of altering F_t you will have to attempt to alter the other side of the equation, the rotating couple between C E and C L R.

The rake of your mast can be changed, or you can move the whole mast forward or aft, although this is not easy if it is taken through the deck. This sort of improvement is drastic and perhaps irreversible, and can also be expensive, so before taking action of any kind you will want to assess the factual situation. Things seldom work out precisely as planned, but simple measurements and calculations will put you on the right path.

First determine the scale of alteration needed. Do you wish, for example, to reduce weather helm by half its angle when your boat is heeled at 18 degrees? Or is it more essential to concentrate on the weather helm experienced under short sail when pressed down to the gunwales? Having defined your problem, consider this diagram.

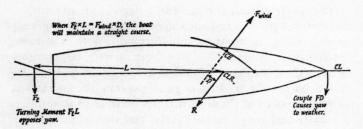

69.   Maintaining a straight course

The weathercocking couple is the product of the wind-force F and the effective distance between CE and CLR, D. For a given windspeed your craft will move at a constant speed and angle of heel. As long as the force FD is balanced by the rudder turning moment your boat will sail in a straight line. To keep her sailing like this, you have to apply 8 degrees of weather helm. You wish to reduce this to 4 degrees. For such small angles it is practical to assume that turning force varies directly as angle ø for a given boat speed. The problem then boils down to a matter of halving the force F D.

The difficulty about this is the shifting nature of C E and C L R, as already discussed, but if you take the designed loci as a basis for adjustment, the necessary period of trial and error will be cut down to the minimum. As with so many facets of boats and sailing, you can only start out with what facts you can surmise from the information available, and then try out your conclusions practically.

This is in effect a caution to proceed softly, one step at a time. If you think of moving your mast, it would be well to jury rig it for trial purposes and to leave permanent re-rigging until you are certain of results. You can juggle about with sail areas in many ways, but I would shy away from re-cutting or re-shaping a sail unless I were sure that I was justified in doing so.

If you suffer lee helm, what follows can be taken in the opposite sense, but by far the majority of boat owners will be troubled with excess weather helm. The first step to reducing this would be to consider adding a bowsprit and small jib. The effect of this would be to move the C E forward and have little effect on the angle of heel, because the combined C E would not be raised by any significant amount. I can think of dozens of chaps who have done this with beneficial effects. (Anyone reading the *Practical Boat Owner* will know what I am talking about and, in passing, let me say that this periodical will be found to be of great value to practical people who like to do things for themselves.) It is best to tackle the problem systematically and without expense as far as possible.

Using your plans, measure the real distance D (in inches) between your design C E and C L R. Multiply the figure by the square feet of sail area you use and call the answer F D; it is convenient fiction to be used for the following simple sums. Draw in the position of a small jib set on a pencilled bowsprit stay to the masthead. Make its area about one-third of your working foresail for a start. Using the Appendix, p. 201 calculate the position of the fresh C E and measure the reduced distance D. Multiply it by the increased total sail area. If this product (F D_2) approximates to half F D you are nearly home and dry. If not, you can lengthen or shorten the bowsprit, or alter the area of the jib until you get the answer. Like many things about a boat, if it looks right it will probably be right. It is then a simple thing to rig a jury sprit, cobble up a temporary sail and try matters out.

The same result cannot be arrived at merely by increasing the size of a foresail set on your existing rig. It is the placement of a new stay which permits the C E to go forward to an appreciable degree.

If you cannot, or do not wish to, add sprit and jib, you may be able to change the position of the C E by altering mast rake. Many masts have some aft rake and there is little harm in reducing or eliminating this, but stop short of forward rake. It is ugly and could lift the boom well above horizontal when it will tend to swing outboard under its own weight and require more effort to sheet in. Usually, however, if weather helm is a real bother, no amount of alteration to rake will suffice to cure it. You will have to think about moving the whole mast forward.

Assuming that your sail configuration stays unchanged, the C E will move the precise distance that the mast is moved. If you can restep the mast without moving chainplates and altering the length of rigging wire it will be fairly easy to experiment. A temporary mast step permitting of small increments of movement can be fitted in place of the existing step or tabernacle:

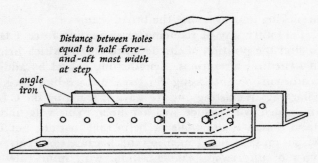

Distance between holes equal to half fore-and-aft mast width at step

angle iron

70. Temporary mast step: mast retained between pair of bolts

Once you have found a suitable position for the mast you can make permanent changes to step and rigging. I would not myself consider re-stepping a through-deck mast; there are too many imponderables such as stress on the hull to be thought about. Even with a deck-stepped mast you should, obviously, make sure that below-deck stiffeners, like beams and pillars, are still able to accept the stresses applied at the new position and distribute them fairly. You might have to shift a pillar or double a beam.

The proportion of sail area carried fore and aft of the mast can be changed in many ways. The C E can be taken forward if you reduce the size of main, increase that of fore canvas, or do both. However, a lot of 'popular' boats are basically undercanvassed to start with and any reduction of this marginal area might seriously affect lightweather performance. Think very hard before cutting a sail down in any way.

One little recognised cause of weather helm is the imperceptibly increasing belly of aging sails, which brings the C E aft. This is applicable especially to mainsails. You can flatten such offenders in a number of ways, like unpicking and re-sewing the broad seams.

The other way of reducing or increasing force F D is to alter the position of the designed C L R, which brings in structural alterations. Lengthening a keel by adding deadwood or increasing fin area aft, adding skeg or enlarging an existing one are ways of bringing the C L R aft, which will reduce weather helm. You will understand that it is the *angle* of helm that matters, not the weight on the tiller. I mention this because some extreme types of underwater configuration, with minuscule fin keel and over-balanced rudder blade, may handle with little weight on the helm but at the same time be carrying far too much angle to the rudder. Beware of structural alterations; they may be expensive and hard to reverse if proved unsatisfactory. It is altogether simpler to concentrate on above-deck modifications.

The additions of dodgers and hoods around the cockpit area brings in much extra windage aft, which can increase weather helm. If adding such weather protection you should try to calculate its likely effect.

12

Roll

The manner in which your boat rolls has a great bearing on performance and also on comfort, which I regard as integral to performance. In hard conditions you need the best from boat and crew together, and heavy motion with insecure footing does not lead to good, safe handling.

In common with other types of motion, roll is brought about by the action of wind and sea, nearly always in combination. It is a sort of oscillatory heeling.

Under sail, the wind acts on the C E to heel a boat, and the angle changes constantly as the wind gusts and lulls. Variations in heeling force (H) impart a rolling gait to your vessel. If she moves from 15 degrees of heel back to 10 degrees, out again to 18 degrees, back to 12 degrees and so on, she is in fact rolling unevenly through an arc on the lee side. Wave action will also be playing a part in moving her.

A boat at rest can be moved away from the vertical by the application of any type of heeling force. When it is removed she will roll back to and then past the vertical. The cycle is repeated until, under the damping effects of air and water (mainly the latter), she comes once more to rest upright.

Any boat can be regarded as having tendencies to be either stiff or tender. A very tender boat will adopt a large angle of heel for a given wind force and this tendency will persist over a great range of angles. She will roll over an unduly wide arc for given sea conditions. A really stiff boat will heel little, but have a rapid, jerky kind of roll which puts great strain on spars and rigging. Some boats seem to adopt an initial fairly steep angle of heel and then get progressively stiffer.

You might consider your boat perfection; I might think her impossibly tender; another chap might say that she was too stiff for comfort. It is an entirely subjective matter but fortunately, therefore, an owner can do a lot to change the behaviour of his boat if he finds it unsatisfactory.

The angle of heel and type of roll experienced for any hull will depend on three main factors: hull section, position of the C G, and the distribution of weight aboard including fixed and movable ballast. The combination of these determines her ability to counter outside forces, her stability in other words. Heeling forces are opposed by transverse stability.

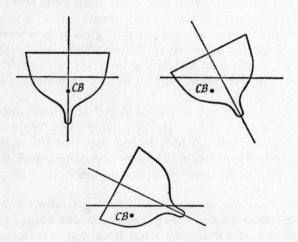

71. Movement of centre of buoyancy, C B

The illustration shows how the C B of a hull moves outboard with heel. You can check on this for your own boat by cutting out shapes of upright and heeled sections and balancing them to locate the altering position of the C B.

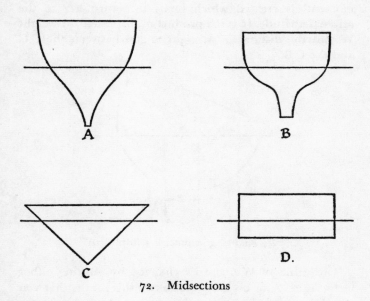

72. Midsections

Midsections are a clear indication of a boat's rolling and heeling tendencies. If you consider those in Fig. 72, you will realise that the slack bilge of A means a small outboard movement of the C B and hard bilge B a rapid one. A will be initially tender and B stiff. Taken to extremes, C will be very tender and D very stiff. A good designer will aim for moderation in his sections and his boats be likely seakindly and good performers.

Unrestrained by outside forces a boat will float so that C G and C B are vertically in line. The total weight (displacement weight: displacement) of a boat acts vertically downward through the C G under the force of gravity. There is an equal and opposite upward reaction at the C B.

If she is displaced from the static position, a 'righting moment' is created which tends to return her to the original attitude. It is the product of displacement weight W and the distance Z at right angles between the C G and the C B.

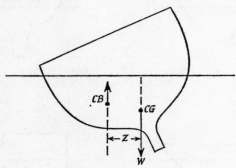

73. Righting moment (simplified)

The value of W Z can be changed by altering either factor W or Z, or both. In real terms this means that you can change your boat's righting moment by increasing or decreasing displacement, lengthening or shortening the righting arm, or lever, Z, or a combination of both. In order to change the vertical position of the C B you would have to alter the hull section, which is out of the question. This means that the only practical way of varying the righting lever is to shift the location of the C G by adding, subtracting, or redistributing weight in the hull. Any changes to total weight will naturally raise or lower the L W L; you can estimate the effect of this from the t.p.i. calculation (see Appendix, p. 200).

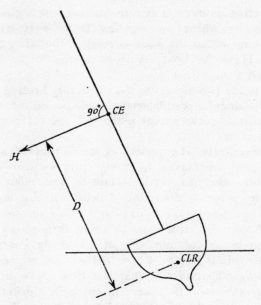

74. Heeling moment = H × D

Heeling force H, which is a component of total wind force F, acts at right angles to the C E. It varies in a manner that means that, for a given wind strength, H is virtually constant up to 8 degrees or so of heel and then becomes progressively smaller. In the limit, no heeling force will be exerted on a sail which is parallel to the water and the wind-flow over it. The heeling moment given to a boat is the product of force H and the distance D between C E and C L R. D is invariable, unlike Z, so that the value of moment H D depends solely on the force H applied at the C E at any moment and similarly becomes smaller as the angle of heel increases.

The situation is, therefore, that as heeling is initiated

the heeling moment is at a maximum and beginning to get smaller whilst the righting moment is zero and increasing. When the stage is reached that they balance $(WZ = HD)$ the boat stabilises at the angle of heel attained at that time.

There are two aspects to the oscillatory heeling known as roll—angular displacement and period of roll. It may help to make things plain if I use the pendulum as an analogy.

Theoretically, the period of oscillation of a pendulum (the time taken to swing from vertical—out to one side—back to vertical—out to the other side—back to vertical) does not depend on the angle to which the pendulum is displaced. This means that it should take the same time to swing through an arc of 10 degrees as through one of 30 or 85 degrees. In practice this is not so. Pivot friction and air resistance (damping effect) dictate that the time taken is longer with greater arcs. If it were to swing in thicker media such as water, the effect would be exaggerated and the difference between times of travel for smaller and larger arcs be more significant.

You can think of your boat (but only for illustration) as a sort of pendulum with its weight, of course, at the C G. There is, in fact, no fixed axis of roll for a boat but it is convenient to assume that she will rotate about the longitudinal axis of flotation. The damping effect of water means that a tender boat, because it rolls through larger arcs, will roll more slowly than a stiff one. The damping works both ways so that the tender boat will take longer to reach the state of heeled stability where $WZ = HD$. She will obviously then be at a greater angle of heel and the reverse roll will be similarly slower.

The amount of damping provided by the water will

depend on the volume displaced as the underwater lateral area of the hull sweeps through an arc, and on the size of that arc also. The size and shape of the area immersed are critical to damping, and there is a direct analogy to be drawn from the action of a rudder being moved through the water (see page 150).

One way to tame the whippy roll of an over-stiff craft is to increase damping by enlarging the immersed lateral surfaces. Bilge keels are often used for their anti-rolling properties. Any such additions should be of lightweight material; heavy stuff could lower the C G, increase W Z, and so have a reverse and undesirable effect. Should you add extra area, it must be in equal amount fore and aft of the C L R if that is to remain unchanged in location.

This brings out the important point that the C G of a boat does not necessarily, in fact not very often, lie below the C B. For reasons of simplicity I may have implied this up to now, but when considering changes in ballast weight you should know of the potential hazards. The matter is discussed in the appendix (p. 205) for the benefit of any reader who owns a small cruiser of the beamy, shallow-draught type such as a centre-boarder or twin-keeler. The C G of such a craft may be quite high and the alteration of weight needs careful forethought.

As the C G cannot be set low in such designs, it is usual to give them extra beam to increase the Z component of the righting moment. This tends to make them initially stiff, but it would be wrong to attribute this tendency to undue weight of ballast, and to remove some. It could result in the boat having negative stability at large angles of heel; she would not be self-righting past a certain point. Quite a few such boats are marginally stable at large angles of heel but as the force H falls off at large angles, you could safely add extra sail area to overcome initial stiffness; this would mean reefing earlier. When reading about leeway later on, you will find that extremely shallow-draught boats have other disadvantages. All in all they are, with the exception of craft over about 35 feet L O A, which despite their light draught can carry adequate outside ballast, not really suitable for off-shore work.

It should be clear, as far as more normal hulls are concerned, that the most practicable way of modifying stiffness or tenderness is by attention to ballast. To cure stiffness, you will have to reposition movable weights to lie higher; add inside ballast; or remove fixed ballast, preferably from as low down as possible, perhaps by taking out a portion of keel ballast and replacing it with deadwood. Adding weight will affect the L W L, and it is often possible to get good results merely by repositioning gear. Watertanks can, for example, often be removed from the bilges and re-sited under bunks, or even outboard of settee backs.

Water ballast tanks along the sides of the hull are a very cheap and effective way of adjusting transverse and longitudinal stability, provided their contents are not permitted to surge; they need to be filled almost to brimming and also have plenty of baffles fitted internally. The famous Eric Tabarly fitted such tanks to one of the *Pen Duicks* and by filling the weather tank only was enabled to make his shallow, 40-foot cruiser plane for days on end over the Pacific. You may not have to resort to such dodges; it is astonishing what a difference it makes to take a 30-lb anchor out of the bilge and chock it on foredeck or even coachroof.

A tender boat is stiffened up by reversing the process. All heavy gear should be stowed low down and you could consider an addition to the weight of the fixed keel, or fin.

I have seen this done with good effect on several boats, and the drawing shows alternative ways of achieving the desired addition. A friend of mine had a *Corribee* and thought it a little tender for his taste. He cast two

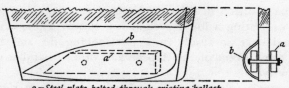

a—*Steel plate bolted through existing ballast*
b—*Glassfibre fairing bonded on*

A *Increasing the weight of a bilge keel*

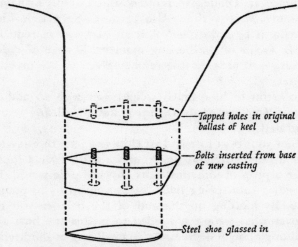

—*Tapped holes in original ballast of keel*

—*Bolts inserted from base of new casting*

—*Steel shoe glassed in*

B *Adding weight to the base of a fin keel*

75. Increasing ballast

hundredweight of lead into a shape to fair on the bottom of the fin keel, bolted it in place and glassed the whole thing over, at the same time incorporating a stout steel rubber along the base of the casting. It not only cured the tenderness, but gave some added weatherliness to the little boat due to the increased draught and A R of the fin.

The addition of fillets of lead, or cast iron, is a good way of getting a lot of weight low down. These can be glassed over to result in a nice-looking fin and bulb keel reminiscent of the Arpege and other desirable craft of the type.

Do not overlook the effect of heavy mast, spars, rigging and fittings. Due to their distance from the mass of the hull, they have a disproportionate capsizing moment which takes quite a bit of the effective righting moment to counter it. Tenderness is often caused simply by having too much weight aloft. Also because the C E is high, lowering it by reducing A R is an easy way of reducing the 'D' factor of the heeling moment. Beware of reducing total sail area for the reasons given in the previous chapter.

To estimate how much weight you wish to add, or remove, you can easily rig up a static inclining test as illustrated.

Moor your boat fore and aft close to a jetty or quayside. Take a line at right-angles from the mast, over a sheave set on a prop of some sort (any chocked pole would do), and down to a spring balance reading up to 20 pounds or so. By hauling on the end of the balance you can discover what weight is needed to incline the boat at a small angle, say 6 degrees. As at small angles the heeling force is almost unaffected by the angle of heel, you can ascertain what adjustments to weight are needed to increase or decrease the righting moment to the desired

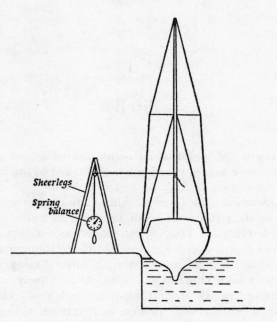

Sheerlegs

Spring balance

76. Inclining test

degree. This will obviate the need for frequent sailing trials for the same purpose.

To make sure that the boat inclines regularly to exactly the same angle, wedge a spirit level on deck so that the bubble is centred at the angle decided on. Differences of but a few minutes of inclination will be easy to see.

Keel weights can be temporarily attached with the boat dried out against a wall. One reading of the balance should show to what extent the weight needs altering before being permanently affixed, but it is as well to have another test to confirm your conclusions before finally attaching supplementary weight.

13

Pitch

Pitching is the longitudinal oscillation of a boat about a thwartwise axis of rotation, usually taken as the lateral axis of flotation (L A F).

Boats which have a bad pitching motion will either dig deep or slam into seas, both resulting in more or less rapid deceleration. Digging can make water stream along the deck and is uncomfortable, but bad slamming is disconcerting; it is at its worst with boats having a 'U' section or hard chine at the bows. Alternatively, a boat will move rapidly up and down through small arcs, or 'hobbyhorse', and this motion is irritating, tiring and can induce sickness. The two extremes of motion can be regarded as stiffness or tenderness in the pitching plane.

The action of waves in causing a boat to pitch is obvious. The passing of a sea lifts her first at one end and then at the other. The way in which your own boat reacts to this depends in great measure on the way in which weight is distributed within the hull. Once again, the analogy of a pendulum will help to make matters clear.

Long pendulums swing slower than shorter ones, irrespective of weight; it is only the position of the C G that matters and the bob is much heavier than the arm, so that the C G is well down in the assembly. So with a boat. Weight which is spread out well to fore and aft can be likened to a pendulum with a long arm and will have a slow rate of pitch; if it is concentrated about the area of the C G the boat will hobbyhorse. This is, however, not quite the whole story because the displacement weight is relevant to the manner of pitching.

The inertia of an object is its resistance to a change in velocity. Inertia depends on weight and speed. You can catch a cricket ball but not a bullet because of the difference in their speeds. You can stop a cricket ball with your foot, but would be unsuccessful (and unwise) to try to stop a steamroller in that way due to the difference in weight (mass).

The moment of inertia of a pitching boat depends on the C G of the two halves of her mass fore and aft of the axis of rotation and the length of the lever on which they act.

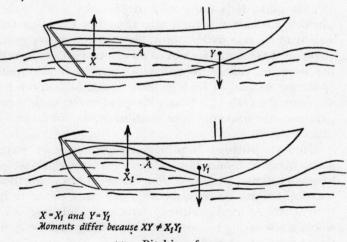

$X = X_1$ and $Y = Y_1$
Moments differ because $XY \neq X_1Y_1$

77. Pitching forces

It is clear from the diagram that a boat with a greater proportion of her weight near the ends will have a greater moment of inertia than one with weights concentrated more centrally. She will move relatively more slowly when pitching, but take a lot of stopping and reversing, whereas the other one will react quickly to outside forces. One will dig or slam, and the other hobbyhorse, or bounce.

Light boats are naturally livelier than ones of the same dimensions but greater displacement because of the difference in mass. You won't be able to do much about this. It is not very practical to add larger weights in an effort to modify pitching tendencies because of the resulting lowering of the L W L with consequent impaired performance. For serious cruising, it is probably preferable to sail in a boat of moderately heavy displacement and accept inferior light air performance although this can, often, be improved by hanging out more washing.

That aside, it is fairly easy to do quite a bit about pitching without structural alteration like changing the weight of outside ballast. Movable stuff like heavy gear, chain, anchors, batteries and so on can be re-stowed until the required change in behaviour has been achieved. Moderate weights of lead or other inside ballast can be added at the ends of a boat with great effect; once their permanent position has been established, do not forget to immobilise them.

Although pitching is principally the result of wave action it can be induced by rolling. This interrelationship between roll and pitch is seldom discussed but is worth a mention because some boats are much more prone to it than others. As hard weather directly causes pitching, it is worth while trying to cure excessive rolling tendencies if your boat is so affected.

Fig. 9 has shown the shape of a heeled waterplane. You can confirm for yourself that if a hull has pronouncedly full after sections the effect of heeling (rolling) will be to bring the L A F aft of the designed position. As the volume of water displaced by the hull must always remain equal fore and aft of the L A F, the result of heeling is that the stern rises and the bows dip in order to equalise volumes. This is the initiation of pitching. When the boat rolls through upright to the other cant, a complete pitching cycle results.

Not only does that happen, but the C L R moves forward as the bow dips and this increases the weathercocking couple. A helmsman will put on weather helm to counter this, take it off as the boat rolls more upright, and re-apply it as she rolls to the other side. A tendency to over-correct will cause excessive yawing and, by so taking the boat into and away from the wind to an unnecessary degree, increase and decrease the heeling moment to such an extent that rolling brought about by wave action is amplified. It can, as explained, also be liable to increase pitching. Such unhandy helming can seriously affect performance and needs to be watched for at all times. A well-balanced hull will need very little major correction when rolling, if any, and will maintain a regular, corkscrewing mean course.

Many old boats, like pilot cutters and other working craft, are still afloat and good for many years yet. Most of them had very fine entries and consequent innate weathercocking potential. It was kept under control by the use of extra foresail area set on a reefing bowsprit which could be adjusted to suit prevailing conditions. A lot of these ex-gaff or lug-sailed boats have been 'modernised' by rigging them as bermudian sloops. This brings out their undesirable rooting tendencies, and the cure for excess weather helm, as previously discussed, is to adjust the C E by means of fore canvas.

14

Other Motion

For the sake of completeness, there are three other forms of motion which I should mention: heaving, surging and drifting. There is also harmonic rolling. What, if anything, you can do to counteract their effect on performance will be obvious from the text and is simply a matter of boat-handling.

HEAVING

Heaving, or tossing, is the vertically up-and-down bodily movement of a boat under the force of seas. Due to inertia, a boat will tend to go on descending into the water when she has been lifted and the wave drops away beneath her. Before doing this, however, she will 'hang' for a moment before gravity gets hold of her and annuls her upward movement. This is an unpleasant form of motion because one suffers a regular reversal of increasing and decreasing 'G' forces which is highly conducive to seasickness.

Once again, light boats will be bouncy and heavy ones stately in their motion. The resistance to inertial movement is provided by the damping effect of the water, and the way in which it takes effect rests with hull shape.

With slack bilges, when the hull sections approximate to a 'V', water will be displaced in gradually increasing increments as the hull descends into it. Damping will

thus take place more progressively and over a longer
period than with a hard-bilged hull when the effect will
be intense and short-lived. This is really only true when
a boat is in a more or less upright attitude as she heaves,
which may not be the case in bad conditions.

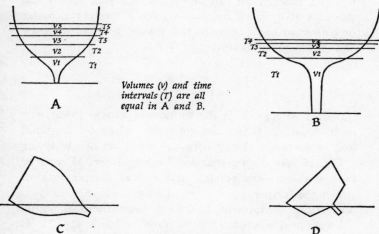

*Volumes (v) and time
intervals (T) are all
equal in A and B.*

78. Bilge shape effects

At 78C a slack-bilged hull is descending at a steep angle
of heel and the same thing is happening to a hard-
chined example at 78D. In this case the latter section
proves advantageous in presenting a better form of entry.
Reaction to heaving is thus to be seen as somewhat unpredictable and dependent on sea conditions. However,
there are steps which you can take to ameliorate them by
controlling your boat's attitude to give the kindest entry.

A slack hull will descend more deeply under most
conditions. If it has full after sections, the inevitable
result will be to start up a pitching reaction.

SURGING

This is a fore and aft oscillation not frequently met with in sailing craft, although it does occur with following seas when a boat alternately surfs on crests and slows in troughs. Power craft can be driven at higher speeds and this may give rise to regular surging. A moderate change of course to meet or cross the seas at a different angle can be beneficial at times.

DRIFTING

Drifting, or leeway, is the motion of a boat thwartwise to its heading. It is most noticeable when on the wind and is generally absent with the wind aft of the beam.

Fig. 26 has shown that the heeling force, H, can be broken down into vertical and lateral components, of which the former may be ignored for practical purposes. The lateral component, L, tries to move the boat bodily sideways and is resisted by the water acting at the C L R. The amount of resistance offered is clearly proportional to the immersed area of the hull, which explains why a beamy, shallow-draught boat is not efficient to windward. A centreboard will increase lateral area when it is lowered and so resist drift. Too much draught can have an adverse effect by lowering the C L R so that the resistance of the water acts on a long lever and promotes heeling.

The subject is complex and with fixed keel boats it is, as I have said before, as well to avoid structural alteration as far as possible. A few additional inches to the bottom of a fin keel might enable a boat to point higher, but with the possibility of induced drag and extra heeling moment being introduced speed could suffer. It is really only practicable to measure the effect of proposed alterations empirically, preferably with a model in a water tank.

What you certainly can do, once again, is to keep driving forces as high, and heeling and drifting forces as low, as feasible by dint of good sail trim and boat handling. Clearly the less H the less L.

Leeway can be measured directly as the angle between course and heading.

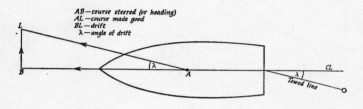

AB—course steered (or heading)
AL—course made good
BL—drift
λ—angle of drift

79. Leeway, or drift

If you sail on vector **A B** and have drift along vector **B L**, your actual course will be along **A L** and the angle of drift, or leeway, angle λ. A lightly weighted line towed from the middle of the transom or counter, which should be long enough to clear eddies and disturbed wake water, will instantly show up the difference between course and heading. By watching your speedometer as you trim sail you should be able to combine optimum forward speed with minimum leeway.

HARMONIC ROLLING

Fig. 80 demonstrates, in exaggerated form, how a boat rolls in a beam sea.

As she starts to climb the oncoming slope her **C G** will start to align itself with the **C B** which has moved outboard because the waterplane is no longer horizontal.

She rolls out from the face of the wave. As she mounts the crest and slides down the reverse face of the wave she rolls through upright and out to the other side, upright again in the trough, and so she goes on rolling.

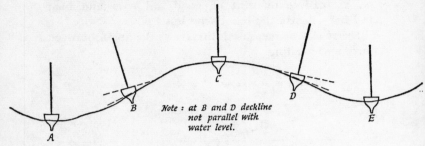

Note : at B and D deckline not parallel with water level.

80. Rolling in a beam sea

Under some circumstances the wave period (time between crests) harmonises with the boat's natural period of roll, and this causes it to build up instead of damping out. If you find yourself in this most uncomfortable and potentially hazardous situation, a change of course to break up the pattern will remedy matters immediately. You can, incidentally, get harmonic, or forced, pitching but this is unlikely with smaller boats. Paradoxically, the more stable a boat is, the worse she can roll in some conditions because she tends to be quick in her reactions to being displaced from the stable; her harmonic roll will be of shorter period than a more tender boat with a slow roll. A stiffish boat will harmonise better with short, steep seas than with longer, smoother ones. Do not confuse this type of rolling with the rhythmic rolling before the wind.

15

Performance Under Shortened Sail

I am going to assume that you have made your own adjustments to performance in the light of what has been discussed so far, and that your boat will attain her best speed in a certain strength of wind under full working sail, usually taken as full main and No. 1 jib. The wind speed will depend on the angle of heel that you have decided is acceptable for comfort and which you have achieved by making corrections to stiffness or tenderness, as suggested in Chapter 12. Some skippers like to reach this state at the top end of Force 3 or thereabouts, while others like to hang on to full sail to the top of Force 4, or over. It is a matter of personal choice, and pleasure.

However, over the chosen limit of wind speed the angle of heel will steepen without any increase in boat speed and the time has arrived to shorten sail. A smaller headsail is set and the main reefed progressively as the wind strengthens. At some point in time it will become apparent that the boat has reached the limit of her ability to make to windward, which is what essentially matters.

There have been many reams written about 'clawing off a lee shore' and the expression is a cliché. Nonetheless, one day you may be forced to beat out of danger and it is essential to know how your boat would be likely to perform in such straits. That will depend on her power to carry sail (see Appendix 217).

This power can be expressed as a mathematical form with reasonable accuracy, but its derivation is a complicated affair for two reasons. It is a laborious matter precisely to locate the C G of a boat; the actual pressure of the wind exerted at the C E depends on the efficiency of the sails in converting wind force to driving force, and can only be approximated.

The sail-carrying properties of boats are compared as ratios of sail area (S A) and displacement weight (W), thus:

$$R = \frac{S\,A \text{ in square feet}}{W^{\frac{2}{3}} \text{ in tons}}$$

The ratio for a Class 1 ocean racer would be around 180-190, similar to that of a 6-metre, some of which have been converted to very able cruisers, albeit a little narrow-gutted. The old J-class boats had a ratio of over 300 which accounts for their ability to exceed the theoretical maximum hull speed, sometimes reaching a value of $1.45\sqrt{L\,W\,L}$. A cruising yacht with the ability to cope with hard weather would have a ratio of no more than 150. This would mean poor performance in light airs but, as I have said, there are many ways of adding effective extra sail area.

As a point of interest, the following table shows the ratio for boats of various displacements, all designed to carry 250 square feet of working sail:

W(tons)	R
1·00	250
1·25	215
1·50	191
1·75	172
2·00	157
2·50	136

Such comparisons are interesting, but the value of the ratio is not really a good indication of a boat's power to carry sail. The displacement factor is for weight alone, and takes no cognisance of its disposition which is critical to stability, particularly in the rolling plane. It merely assumes that boats of similar weights will have similar C G positions, or similar righting moments; if not, then that the angle of heel is immaterial but, for purposes of both comfort and performance, there is a rational limit to which you can press a boat, usually no further than lee rail awash. After this form drag begins seriously to effect performance.

The assumption may have been more valid in the days when boats were uniformly built of timber and did not vary widely in their hull characteristics. This would be up until new building materials were introduced, such as resinglass with its high strength-to-weight ratio, and also extreme underwater profiles like deep, narrow fins and high A R rudders.

For example, consider the case of an anonymous but well-known 'popular' 22-foot sloop. She is produced in both fin- and twin-bilge-keel models which have the same sail area, identical displacement and ballast weights and, as the hulls all come from the same mould, the C B is only marginally affected by the difference in keel volume. However, the C G of the fin-keel model is roughly 10 inches lower than the other.

It is apparent that the righting moment will be quite different for each model, and that the boat with the shallower draught will not be able to stand up to her canvas as well as the finkeeler. This makes a nonsense of the accepted ratio. One model must obviously be either over- or under-canvassed in comparison with the other.

Alternatively, there could be boats identical in sail area, displacement and height of C G but which had differing bilge forms. Shifts of C B would mean that righting arms would differ, and so would the righting moments. The boats could be equal in ability but would vary in stiffness.

Even if there were an easy theoretical way of finding out your boat's power to carry sail, it would be of little practical value for assessing probable performance under deeply reefed conditions, as will transpire. Consider the comparative requirements of two 200-square-footers differing only in the height of their C Gs.

One sails at best speed at 15 degrees angle of heel in a wind of 10 knots; the other does the same in a 15-knot breeze. If the wind increases to 25 knots, what amount of canvas can be carried by each boat if the angle of heel remains the same?

The heeling moment for a sailing boat is given by a formula (which is only brought in momentarily and can then be forgotten) thus:

$$M_h = P \times SA \times D \times \cos^2\theta \text{ ft-lb}$$

where P is horizontal wind pressure in lb/sq. ft., S A is sail area in square feet, D is distance C E-C L R and θ is the angle of heel. S A and P are the only terms we need to bother with, as D and θ remain constant. P $\times$ S A is the total wind pressure acting on the sails; as long as this does not alter, the angle of heel will remain the same.

For 10 knots of wind speed P=0·34 lb/sq. ft., for 15 knots P=0·76 lb/sq. ft. and for 25 knots P=2·12 lb/sq. ft. The steep increase is because pressure varies as the square of wind speed. For the 10-knot boat, the sail area to be carried in 25-knot winds would be $\dfrac{200 \times 0\cdot34}{2\cdot12} = 32$ sq. ft.;

for the 15-knot boat, $\dfrac{200 \times 0\cdot76}{2\cdot12} = 72$ sq. ft.

It is necessary to make a rough adjustment to these

figures because the C E will drop as sail is reefed and shorten the lever D. If it drops by 3 feet for the first boat and 2 feet for the second, D originally being 14 feet, the area required will be 41 sq. ft. and 84 sq. ft. respectively.

These amounts of sail are too small, that for boat 1 being grossly inadequate, but the point has been made that a boat which is stiffer, and can carry the same amount of sail for the same angle of heel in stronger winds, will be able to carry relatively more sail in the reefed condition.

You might think that by using the calculation for heeling moment in respect of your own boat, you could arrive at the best area of reefed canvas to carry for given wind strengths, and determine how much heel they would impose on your craft. This is not so because imponderables enter into the picture.

The way in which sails convert the total wind force into driving, heeling and drifting components depends on camber, interference with air flow, aspect ratio and other attributes of and influences on the reefed sails.

In hard winds it is essential that camber be very small, and the effective wind pressure on the sails will be proportionately less than if the camber were greater as is usual for lighter airs. This means that more sail area can be carried if it is well flattened; I pointed out on page 115 that sails are cut with less camber as they decrease in size. The way in which the total reefed area is divided between fore and aft sails needs consideration.

The mainsail will be baffled by the presence of the mast, and in a blow the interference on the lee side is gross, so that most of the drive comes from the weather side. The mainsail becomes relatively much less efficient than the foresail, and this is where you have to be careful not to let the C E get too far forward and start to introduce lee helm. (See Chapter 3.) Falling off and perhaps broaching could have unfortunate consequences under the prevailing conditions. The foresail area to be carried will be less in proportion to after canvas than in lighter conditions. It is quite surprising what an effect just a scrap of a jib has on steering and handling in a blow.

In practice, I would seldom reduce the mainsail to less than three-fifths of its full size, and never to less than half. Less area than this would reduce the drive/drift ratio to an unacceptable degree, and take the C E too far forward for balance.

The storm jib should be about one third of the size of the working foresail, and its contour needs thinking about, as does its sheeting.

Despite its smallness, the jib will provide much more driving force per square foot than will the main, as its luff is clearer. In beating to windward, you need to take advantage of this fact and this argues that the foresail luff should be as long as possible, in other words a sail of high A R is wanted. This tends to take the C E slightly forward in comparison with one of lower A R, but with a very small sail it is of small consequence.

In bad weather there is a lot of water flying about, and it is essential not to allow any quantity of the green variety to dump itself into the belly of the foresail because it will split it or tear out the clew cringle. This means keeping the foot high up out of the way, so that the ideal sail will be of more or less equilateral shape, as in Fig. 81.

The slight raising of the C E can be ignored.

Sheets should be so positioned that the jib can be sheeted hard without backwinding the main and further reducing its efficiency; they will be more outboard than those of the working jib. There will be little opportunity

81. Storm sail

to ease them if they are set too far inboard, as the sail will start to flog as soon as it is relieved of tension.

It is regrettable, but hardly surprising, that there is very little to be found written on the subject of the effectiveness of a deeply reefed rig. It becomes a matter of trial and error to find the sail area suitable for your boat to be able to make to windward in the hardest weather she can take before lying-to. Once more, the experience of good designers will ensure that the reefed measures designed into their sail plans will be as good as can be desired.

A point arises here about unfair mast loading. It is sensible to reef so that areas are pretty well balanced fore and aft not only because of C E considerations but because of the possibility of the mast breaking.

Changing foresails in bad weather is arduous work and there is always a tendency for a crew to visit the plunging foredeck as infrequently as possible. Thus, one often finds a boat with a well-reefed mainsail and an unduly large masthead sail left set forward. The total sail area may be small enough for the circumstances, but look at Fig. 82.

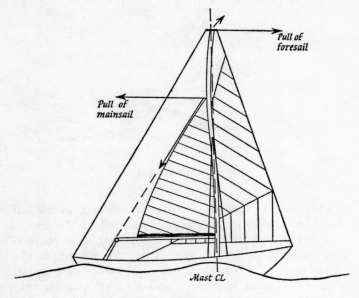

82. Unfair mast loading

The loading due to the mainsail is concentrated on a shorter length of mast than if it were unreefed. This is inevitable, but on the wind the sail will be strapped hard down, presumably, and a load applied through the leach to a point on the mast between crosstrees and truck. This point lies distant from support from either upper or lower shrouds and there will be a tendency for it to bend aft.

The forward pull of the large foresail increases this effect so that the mast assumes an S-curve, as shown. It may then fracture from compression applied to a distorted tube.

A similar unfair stress is applied in a thwartwise direction at the same time. It is probably the combination of the two loadings that is the cause of frequent mast failure when areas fore and aft are unbalanced.

Beyond the point when you can no longer make to windward it is pointless discussing performance. You enter the ambit of survival conditions which need a book to themselves.

APPENDIX I

Apparent Wind

A moving boat experiences a wind coming from a different direction, at a different strength, than she would if stationary. This wind is the only one that matters for practical purposes and is the apparent wind, or simply 'the wind'; the true wind blowing over the sea in the vicinity is either called by that name or known as ambient wind.

The strength and direction of the apparent wind of the moment depend on boat speed and the angle she makes with the true wind, which is of academic interest.

As true wind speed increases with height, the apparent wind blowing on the sails will be at different speeds and directions at different levels (see page 110).

When a boat has the wind dead astern, true wind and apparent wind blow from the same direction, but apparent wind speed will equal true wind speed minus boat speed. This is the well-known snare for the inexperienced or incautious. A boat running at 5 knots before a 10-knot breeze will feel a 5-knot apparent wind, little more than a zephyr. If she then turns on to a closehauled course, the apparent wind speed will rise to 15 knots. As explained in Chapter 15, wind pressure increases as the square of its speed, so that the increase in wind force will be 9 times.

The vector diagrams show a true wind of 10 knots and a boat speed of 5 knots, the angle between course and apparent wind and its strength for attitude of beating, reaching and with the wind over the quarter.

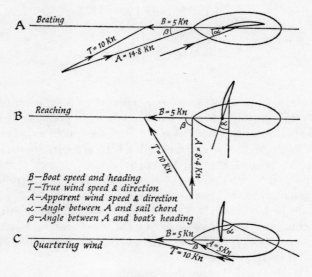

A Beating
B Reaching
C Quartering wind

B—Boat speed and heading
T—True wind speed & direction
A—Apparent wind speed & direction
∝—Angle between A and sail chord
β—Angle between A and boat's heading

83. True and apparent wind

APPENDIX II

Aspect Ratio

The aspect ratio of a sail is expressed as a relationship between its height and width. For this purpose it is always assumed that a sail is flat, and the width expressed as the length of the mean chord, which is found by dividing the area of the sail by its height.

Sails are often drawn out as triangles for purposes of calculation, but to determine accurately their A R you would be better advised to measure their full area with a planimeter. This is essential in the case of sails having great roach to their leach, such as some carried on multi-hulled craft.

To obtain A R, use the formula

$$A R = H \div \frac{SA}{H} \quad \text{or} \quad A R = \frac{H^2 \ (\text{ft.})}{S A \ (\text{sq. ft.})}$$

APPENDIX III

To Find the Displacement
of a Boat

A boat displaces a volume of water equal to the volume of the immersed part of the hull, i.e. all below the L W L. As 35 cubic feet of water weigh one ton, if you divide the immersed volume by that figure you will arrive at the displacement weight expressed in tons.

The displaced volume is found by using measurement with a planimeter and applying Simpson's Rules to the results. This is a simple means of performing the mystical mathematical process known as integration. If you follow the rules, the sums will give as good results as could possibly be expected.

1. Measure the area of each section below the water-line as given on the body plan. Remember that the plan is halved, so that you can only measure half a section at a time and will have to follow the contour twice around to get the right answer. Don't fall into the trap of following a line across the mid-line.

2. Scale up your measurements to give the areas in square feet. This is straightforward; one square inch on a plan scaled at 1 in. : 1 ft. will equal one square foot. An area of one square inch on a plan scaled at $\frac{1}{2}$ in. : 1 ft. will equal four square feet.

3. Construct a table as follows:

Station No.	Area of section at that station	Multiply by	Product
0	0·00 sq. ft.	1	0·00
1	0·60	4	2·40
2	2·10	2	4·20
3	5·34	4	21·36
4	6·30	2	12·60
5	8·12	4	32·48
6	8·20	2	16·40
7	6·62	4	26·48
8	5·00	2	10·00
9	1·22	4	4·88
10	0·00	1	0·00
		Total	130·80

There are a couple of points to note. There must always be an odd number of stations, or an even number of divisions, to put it another way. The multipliers must run in the progression 1 ... 4 ... 2 ... 4 ... 1. As stations 0 and 10 end on the L W L there are no areas to be measured there but you should include them in the calculation if there are. I would prefer to re-section the L W L in such a case for simplicity of working.

4. Multiply the total by the distance, expressed in feet and decimals of a foot, between stations and divide the result by 3; this gives the displaced volume in cubic feet:

$$\frac{130 \cdot 80 \times 2 \cdot 3 \text{ (say)}}{3} = 100 \cdot 28 \text{ cu. ft.}$$

5. Divide by 35 and you have the displacement in tons:

$$\frac{100 \cdot 28}{35} = 2 \cdot 865 \text{ tons}$$

For fresh water, use a divisor of 36 instead of 35.

APPENDIX IV

To Find the Centre of Buoyancy

Construct a table, using the products derived in the displacement calculations, but with a different set of multipliers, like this:

1.

Station No.	Previous product	Multiply by	Product
0	0·00	0	0·00
1	2·40	1	2·40
2	4·20	2	8·40
3	21·36	3	64·08
4	12·60	4	50·40
5	32·48	5	162·40
6	16·40	6	98·40
7	26·48	7	185·36
8	10·00	8	80·00
9	4·88	9	43·92
10	0·00	10	0·00

First total: 130·80　　　　Second total: 695·36

2. Multiply the second total by the station interval and divide it by the first total:

$$\frac{695 \cdot 36 \times 2 \cdot 3}{130 \cdot 00} = 12 \cdot 227 \text{ ft.}$$

This is the distance of the C B aft of station O. As long as the boat floats level on her L W L, because she is symmetric about her fore and aft C L, the C B must lie vertically below that line. This gives two dimensions and, once you know the distance of the C B below the L W L you will be able to spot it accurately on all three plans and so locate its actual position within the hull.

That distance is arrived at by constructing a table in exactly the same way as in 1 above, but using the areas of waterplanes taken from the half-breadth plan (remembering that they must be gone round twice) and the interval between them; this is usually 6 inches for boats under 24 feet L W L. The first total can also be used to find displacement; multiply it by the interval between waterplanes and then divide by 3×35. This can serve as a useful check on the displacement calculated by using areas of sections.

APPENDIX V

To Find Tons Per Inch Immersion

According to her actual displacement at any given time, a boat will float above or below her D W L and it is useful to know how far she will sink with increasing weights of crew and cargo. A simple sum gives the answer.

Measure the area of the D W L plane as shown on the half-breadth plan and scale it up into square feet; say for example that it is 100 sq. ft. Ignoring the slight differences between waterline planes very close to one another, it is clear that if the boat sinks one inch below her D W L she will then displace an extra $100 \times \frac{1}{12}$ cubic feet of water. In tons weight this will be $100 \times \frac{1}{12} \times \frac{1}{35} = 0\cdot238$ tons (about 533 lb).

Supposing you build from a bare hull and, on launch, find your craft floating two inches above her marks. The calculation will tell you that she needs about half a ton of weight added to bring her down to the D W L. If you think that the weight of crew and gear will be insufficient once she is in cruising trim, you can think about adding inside ballast. If she is grossly light and floats several inches high, mark in her actual L W L on the drawings and measure that plane. Add the D W L plane area and halve the result; use this figure for the t.p.i. calculation.

APPENDIX VI

To Find the Centre of Effort

This is an exercise in geometry and needs no calcula
tion. Using your sail plan, take the mainsail and working
foresail as an example. The foresail will be shown as a
triangle; ignore roach of the mainsail and pencil in a
straight leach. This should touch the after corner of the
headboard and be carried on until it meets the projected
line of luff; the resulting triangle is used for the C E
measurements.

1. Bisect two sides of each triangle (sail) and join
the points of bisection to the opposing angles. Where
they cross is the C E of the sail.

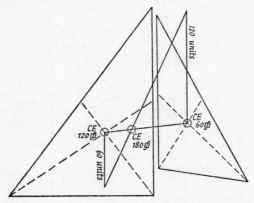

84. Finding the combined centre of effort. C E, of two sails

2. Join the C Es of both sails, as shown.

3. From each C E draw a perpendicular line, one upwards and the other downwards; it does not matter which is which.

4. On each line scale off a length directly proportional to the area of the *other* sail.

5. Join the ends of the measured lengths. Where they intersect the line joining the C Es will be the C E of the sail plan.

If you have multiple sails, or sails which are not triangular in shape (gaff: gunter), treat each pair of triangles at a time until you finally arrive at a single C E, as described. The odd-shaped sails can all be divided into triangular sections for the purpose. Fig. 85 shows how to find the C E of a gaff sail.

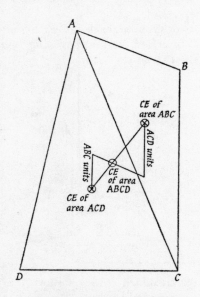

85. Finding the C E of a gaff sail

Sails with excessive roach should be treated a little differently. Cut out their contour in cardboard and balance the shape on a needlepoint. The C E will be in the same place as the C G (see page 28), which is the point of balance, naturally.

APPENDIX VII

The Centre of Pressure
of a Rudder Blade

The C P of a rudder is a moving locus, varying with speed of flow, sectional profile and angle of attack, or incidence to the water flow.

DEPTH OF CP

For a rectangular immersed blade, or portion of blade, it can be assumed that the C P will be two thirds the way down. It will be slightly higher than this if the top edge of the rudder is wider than the bottom edge, and conversely, but this is of no great consequence.

LATERAL POSITION OF CP

For practical purposes of calculation, it may be taken that the C P lies about 40 per cent of the mean blade width abaft the leading edge of a non-balanced blade.

For a balanced blade, it will serve to divide its area into sections fore and aft of the pivot line. Measure the areas. Take the C P of the fore part as being 35 per cent of its mean width from the leading edge, and that of the after part at 40 per cent of its mean width aft of the pivot line. Find the resultant C P by the perpendicular scaling method as for centres of effort.

APPENDIX VIII

Metacentres and
Transverse Stability

A metacentre is a theoretical locus used in stability problems and, like other maritime loci, a shifting point.

Longitudinal stability is comparatively so great that it is not worth bothering about; in normal waters there is little likelihood of a boat pitchpoling, or being capsized end over end. Transverse stability is another matter because excessive roll is quite common.

The transverse metacentre, M, can be regarded as fixed in relation to the hull for angles of heel up to about 8 degrees, and its position can be stated thus:

'M lies at the intersection of two lines on the upright section of the body plan which contains the C G and the C B. One line passes through K, the middle of the keel base, G, the position of the C G and D, the midpoint of the deck. The other is a vertical line passing through B, the position of the C B.'

From the diagram you can see that, when a boat is upright, both lines coincide because C G and C B lie on the line joining K and D (Fig. 86).

If the boat heels, the C G stays at point G because it is fixed with respect to the hull but the C B moves out to point B_1. A line drawn vertically (at right angles to the L W L when heeled) through B_1 cuts the line K G D at M, which is the position of the metacentre.

The simplest way of locating M is by geometry and empiric means. The C B of a heeled boat does not correspond with the C B found by using the upright section containing it because, due to the differing sizes and shapes of heeled sections fore and aft of it, there is a volumetric displacement; one section can only indicate a two-dimensional shift.

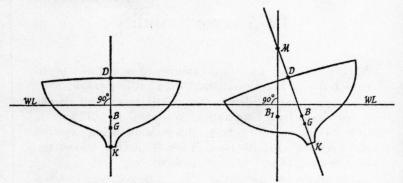

86.   Finding the transverse metacentre

On the body plan draw a fresh L W L to correspond with the small angle of heel chosen. It will cut through all sections. With a scissors, cut out all the fresh immersed sections lying below the heeled L W L. Paste them together so that each L W L and mid-line coincide. Balance this paste-up on a needlepoint and the point of balance will be the heeled C B. Using a drawing like Fig. 1, you can now mark in old and new C Bs and so accurately spot M.

Strictly speaking you should have applied a 'layer correction' to each section to allow for the effect of heel, but this is a fiddling affair and not worth it for the small difference which would be found in the position of the positive C B for any reasonably balanced boat at small angles of heel. However, to make the picture complete for those who may wish to examine the question in respect of greater angles of heel, the diagram shows the effect of heeling on an after section of the body plan.

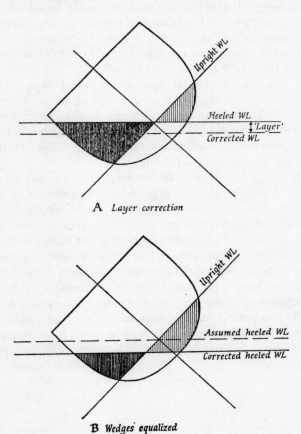

A *Layer correction*

B *Wedges equalized*

87.   Layer correction

Due to hull form the 'immersed wedge' (shaded dark) is greater than the 'emersed, or emerged, wedge' (shaded light). Because the immersed volume of a hull must always be the same, no matter how placed in the water, the immersed and emersed volumes must be equal. To make the immersed area of the section shown equal to the emersed one the L W L has to be adjusted to lie somewhat lower than the one drawn in using the centre of the level L W L as an axis of roll.

It is a matter of trial and error to find it; use your planimeter to measure the areas resulting from changing the L W L position until they are equal. In order to arrive at the true heeled C B, for any angle, you will have to repeat this operation for each section before cutting out, pasting up and balancing.

As the L W L is lowered, or conceivably raised in the forward sections of some designs with fine entries, it will cut the line K G D at a different spot. This means, in effect, that if the immersed area is greater than the emersed one the hull will rise bodily out of the water at that station. With older craft, such as working boats with straight stem and fine entry, the stern used to rise and the bows sink; this caused a 'rooting' or heavy weathergoing tendency with excessive weather helm needed to correct it. Scandinavian boats, with 'pointed' ends fore and aft tend to lift bodily out of the water and have an unpleasant motion.

APPENDIX IX

To Find the Centre of Gravity

You can now calculate the position of the C G as follows:
1. When finding the displacement you measured the areas of all the upright sections lying below the D W L (or L W L); see column 2 of item 3 on page 196. Add all these areas together and multiply the total by 3. In the example this comes to 130.5 square feet.
2. Measure across the top of each section to find the width of the W L at that station. Cube each width separately and add all the cubes together. Say that this total comes to 326.25 cubic feet.

3. Divide answer 2 by answer 1: $\dfrac{326.25}{130.5} = 2\cdot5$ feet.
This is the vertical distance of G below M.

Join G with a line meeting B M at right angles at point Z. The distance G Z is the *righting arm*, or lever, referred to frequently in the chapters. The *righting moment* will, of course, be the product of displacement and righting arm, W × G Z; it is usually expressed in foot-tons but can be in foot-pounds or a metric equivalent if you like.

You will notice that the relative positions of G and B do not matter as long as G lies below M, a condition of positive stability. As long as this is so there will exist a positive righting moment and the boat will always recover from heeling. The further down that G lies in relation to M the longer will be the lever G Z and the greater the righting moment for a given displacement W.

M is, as stated, a movable locus. Once the boat heels more than the small allowable angle below which it is virtually fixed, the metacentre becomes depressed and possibly moved off centre. If it drops to coincide with G the boat will be in a state of neutral stability and have no power of recovery. If the angle is further increased a situation can be reached where M lies below G and stability is negative. This means that the righting moment is applied in the wrong direction and the boat will capsize. This sort of thing is shown vividly by the way in which a dinghy is stable upright when unloaded but capsizes when an incautious crew steps on the gunwale. In this case G has risen to lie above M but the result is inevitably the same—negative stability, heel and capsize.

To counteract this distressing tendency a small boat like that is given great beam. On initial heeling the C B shifts rapidly outboard and the righting arm lengthens sufficiently to provide a moment of considerable size over a small range of angle of heel.

As far as keelboats are concerned, it is fair to say that those carrying a quantity of outside ballast set well down will always have positive stability and self-righting properties up to an angle of roll of 180 degrees: they will recover from a total inversion. This is not necessarily so with vessels of shallow draught and some of them, although having wide beam and initially great stability, may be so designed that there is an angle of heel beyond which they will capsize.

APPENDIX X

Tests of Comparative Stability

The owner of a shallow-draught boat may well wish to know if she is self-righting and it is a simple matter to find out.

By means of a line leading from the masthead (perhaps a halyard) she can be hove down until her truck touches the water. If she staggers back to the upright, no matter how hesitantly, she is obviously self-righting from an angle greater than 90 degrees. She may nevertheless be capsizeable if this angle is only slightly exceeded. Luckily, a combination of wind and seas likely to bring about such a situation is not likely to be met with by the average cruising owner.

Some boats can be capsizeable at smaller angles of heel, say 75 degrees or so, and once hove down beyond this point will roll over. Suspect craft should be tested in shoal water for ease of recovery. It is far easier to approach the testing in easy stages and infer likely tendencies than to try such drastic methods

Once the rail has gone under, the shape and size of decks and upperworks have to be considered because of their effect on buoyancy. As previously explained, the heeled C B may have the effect of bodily raising the hull, so that a boat is heeled so that the L W L lies as shown at Fig. 88A.

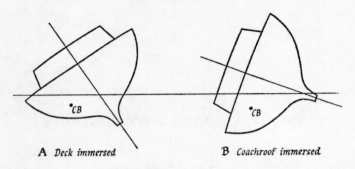

A *Deck immersed*   B *Coachroof immersed*

88.   Shifts of C B

Further heeling brings part of the coachroof down into the oggin; at 90 degrees the immersed volume includes some of the upperworks. As long as these are watertight the boat will ride high, on its side, in the water and the angle of heel at which the truck meets it will be quite a bit over 90 degrees.

Of course, once water gets over the rail of an open or partially decked boat it will enter the bilges with predictable results: the boat will fill and sink.

Assuming watertightness, as upperworks become immersed the C B of a positively stable boat (at that angle) will shift farther away from the C G. Ignoring the position of the now immaterial metacentre, the result of this is to increase the righting ability of the boat due to the greater coupling arm between C B and C G, as shown. In practice this means that a watertight cruiser will seldom be so pressed by the wind that her mast touches the water; the heeling moment is decreasing rapidly with great angles of heel. In big seas, of course, the mast might strike an approaching crest.

The masthead float seen adorning some catamarans and other maritime caravans is a means of giving a great shift to the C B of the assembly once it touches the water. The righting moment becomes greatly in excess of the capsizing one.

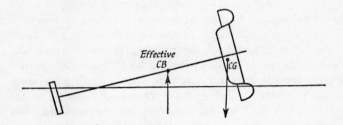

89. Shift of C B with masthead float

In considering stability it is customary to plot righting moment against angle of heel, which gives curves like this:

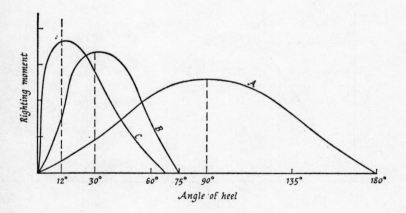

90. Stability curves

Curve A represents the stability characteristics of a boat which is self-righting up to 180 degrees, the case for most deep-keeled boats with heavy outside ballast. Curve B relates to a boat which acquires negative stability over 75 degrees of heel. This might apply to a very shallow hull with inside ballast, or some amount of it. Curve C is for an unballasted, un-masthead-floated catamaran.

At any angle of heel the righting moment equals the heeling moment, so that it would be just as valid to plot angles against this and, as you know, heeling moment is the product of heeling force H and the length of the arm C E-C L R. You can therefore use the inclining test suggested above to find the stability curve of your boat. The weight applied to the masthead will be less than if it were applied to the C E in the proportion found by dividing the distance from masthead to C L R by the length C E-C L R.

By inclining your boat at, say, 10 degrees at a time, you can read off the weight applied by means of a hefty spring balance attached to the line used. Apply the con-

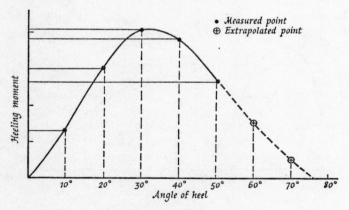

91. Estimating point of neutral stability

version factor mentioned in the previous paragraph to enable you to set heeling (righting) moments on the ordinate and angles of heel along the abscissa. Gradually you will see the stability curve appearing.

Typically, shallow, beamy boats will have their points of maximum stability somewhere about 30 degrees of heel whereas really stable ones will be most stable at an angle of 90 degrees. Multihulls may peak at about 12 degrees.

Beyond the maximum point, the righting moment decreases, and consequently the heeling force needed to achieve a given angle of heel will progressively get less. By extrapolating the curve as shown by the dotted line you can estimate the angle at which your boat will reach a state of neutral stability, i.e. where the curve cuts the abscissa. Heave her down to this angle and you should be able to hold her there with your little finger. A slight shove downwards and she will start to capsize.

Under sailing conditions the heeling force of the wind drops off greatly at large angles of heel. As long as you carry a rational amount of sail for prevailing conditions you need not fear capsize, but you should bear the potential hazard in mind. Harmonic rolling could be fraught.

These considerations apply in exaggerated manner to multihulls. They have enormous beam for their length and extreme initial stability up to around 10 to 12 degrees as shown in Fig. 90. Once over the point of maximum stability they accelerate into a stage of negative stability and then become very firmly stable upside down. I have no wish to sound alarmist, for multihulls are both comfortable and able seaboats. However, they need to be sailed intelligently, and there is always a need to be certain that they are not overcanvassed for the prevailing conditions. An understanding of their potential shortcomings is essential, and you will find that experienced multihull owners are meticulous in attending to their sails. Wave action alone has never, to my knowledge, been responsible for capsizing a multihull; with all sail down they become excellent rafts.

Due to its shape, even a heavy centreboard will not make much difference to the stability of a monohull when it is fully lowered. The modern, movable keel with a heavy bulb at its base is reasonably beneficial in lowering the C G of a lightweight craft. There is, undoubtedly, no real substitute for a good weight of outside ballast, set well down.

APPENDIX XI

The Power to Carry Sail

Within the limits under your control you will doubtless settle on a degree of stiffness or tenderness in your boat which is to taste. However, it may be of interest to have a standard by which you can compare your boat's abilities in comparison with other craft, irrespective of size and type. The sail area/displacement value given on page 184 is not, as explained, of much real benefit.

Obviously, size and weight have a direct bearing on accommodation, comfort and motion, but from the standpoint of seaworthiness there should be little essential difference between well-found craft. It does not matter whether they are of light or heavy displacement; of greatly differing hull form; rigged quite differently; or anything else as long as they all have satisfactory stability characteristics.

A very desirable attribute of a cruising boat is that she shall stand up to her canvas, i.e. not to have to be reefed too early or too frequently. Also, that when reefed she shall continue to be able to make to windward in hard conditions. This ability is referred to as 'the power to carry sail' and can be expressed in various ways. It is a matter of concern to designers, many of whom have evolved a pet way of finding the magic amount of sail area suited to their creations. As their ideas can differ quite significantly, some boats are found to be more able than others in this respect.

There is an index, P, which indicates power to carry sail; being a value, or dimensionless figure, it can be applied to any boat. You find the index like this:

$$P = \frac{W \times GM,}{SA \times D} \text{ where}$$

W is displacement weight in *pounds*

G M is the distance between C G and metacentre

S A is sail area in square feet

D is the distance between C E and C L R in feet.

P should not lie below 3.75, or the boat is too tender, nor above 7.5, or the boat will be too stiff.

With other information already given in this appendix, you are in a position to work out P for your own boat without bother.

List of Recommended Reading

Admiralty *Sailmaker's Handbook* (BR 2176), HMSO
Baader, J. *The Sailing Yacht*, Adlard Coles, 1965
Barnaby, K. C. *Basic Naval Architecture*, Hutchinson, 1967
Bowker & Budd *Make your own Sails*, Macmillan, 1960
Delmar-Morgan, E. *Small Craft Engines and Equipment*, Adlard Coles, 1963
Desoutter, D. *Small Boat Cruising*, Faber, 1964
Fox, Uffa *Sailing Boats*, Newnes, 1966
Howard-Williams, J. *Sails*, Adlard Coles, 1969
Illingworth, J. *Further Offshore*, Adlard Coles, 1971
Marchaj, C. J. *Sailing Theory and Practice*, Adlard Coles, 1964
Teale, J. *Small Boat Design and Construction*, Temple Press, 1964
Warren, N. *Marine Engine Conversions*, Adlard Coles, 1972

The following two books are out of print, but valuable. They can be found in most reference libraries and sometimes copies are available for loan.

Butler, T. Harrison *Cruising Yachts: Design and Performance*, Ross, 1945
Kemp, Dixon *Yacht Architecture*, H. Cox, 1885

Index